To LeRoy,
The man with the magic hands. Thanks for saving my back!
Best,
Arthur
9-06

ARTHUR GRACE

STATE FAIR

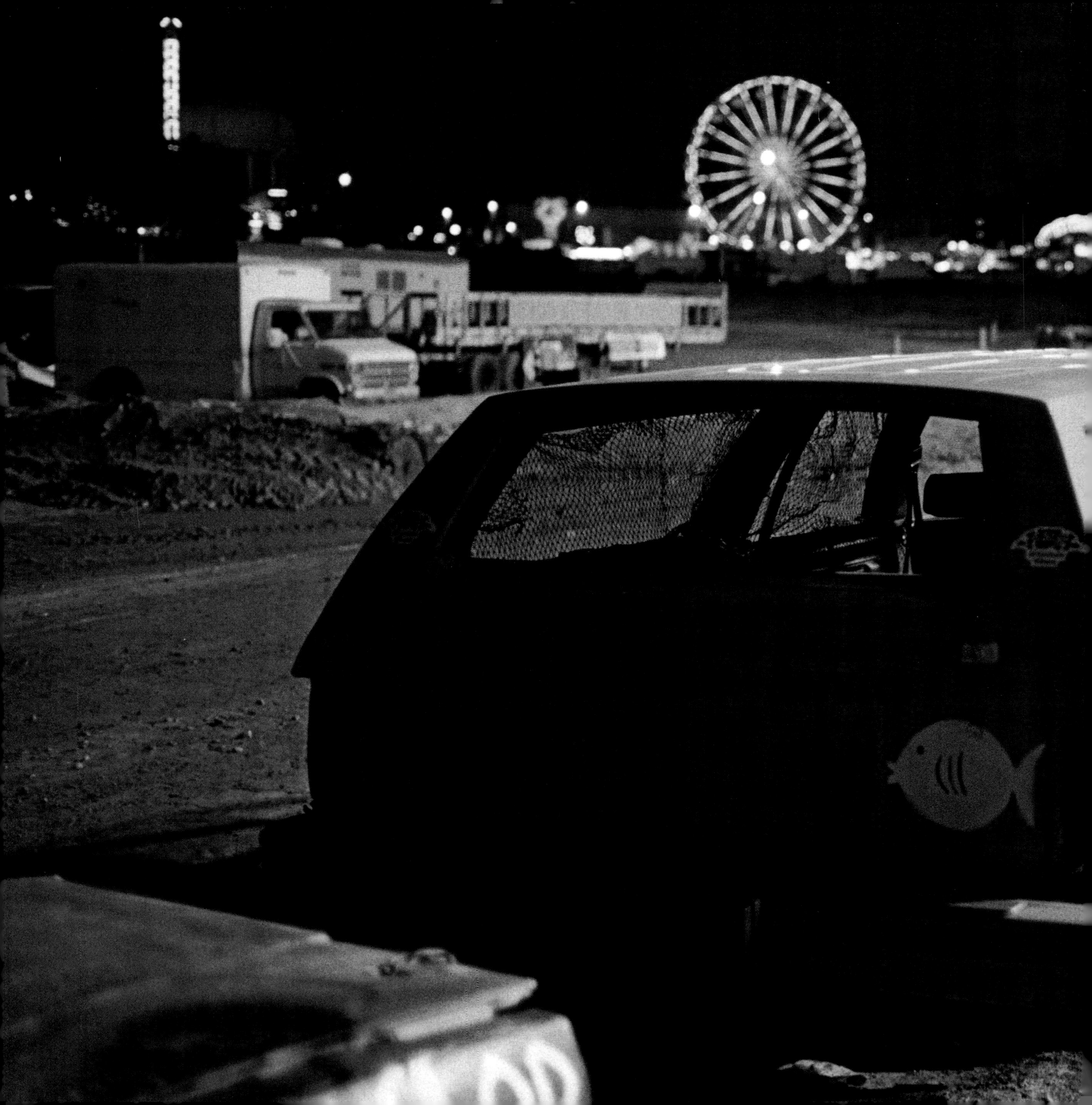

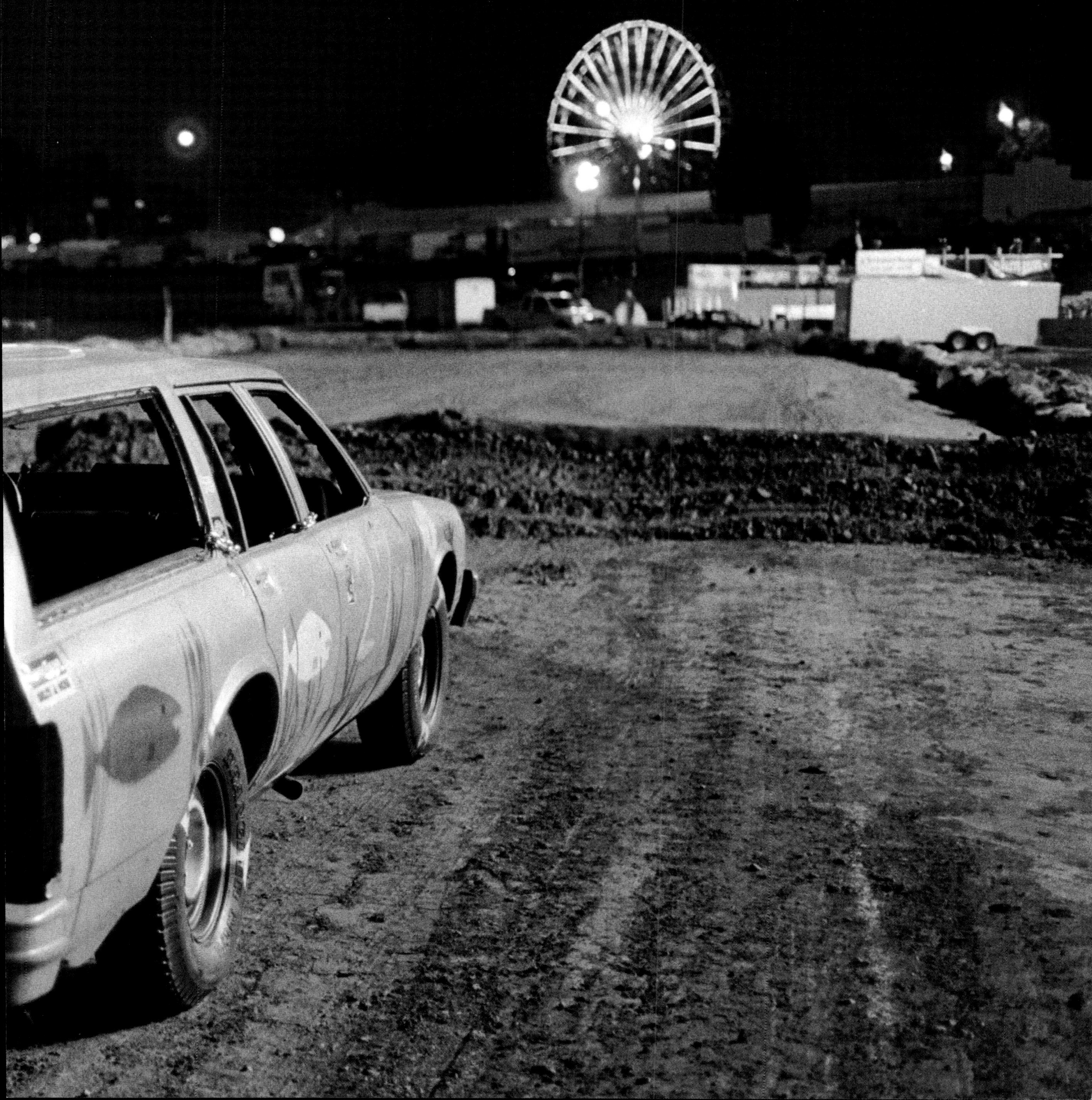

STATE

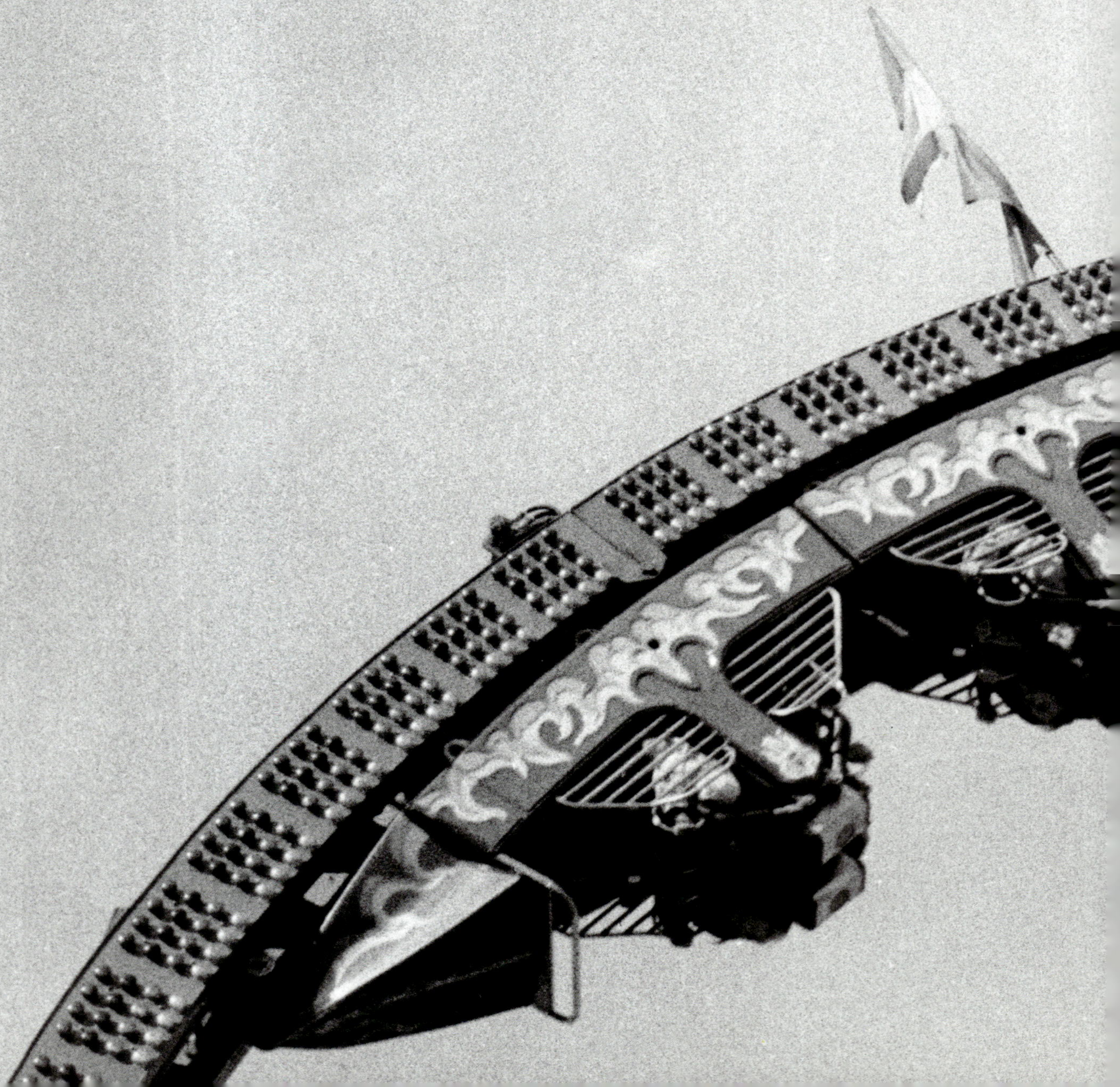

FAIR

ARTHUR GRACE

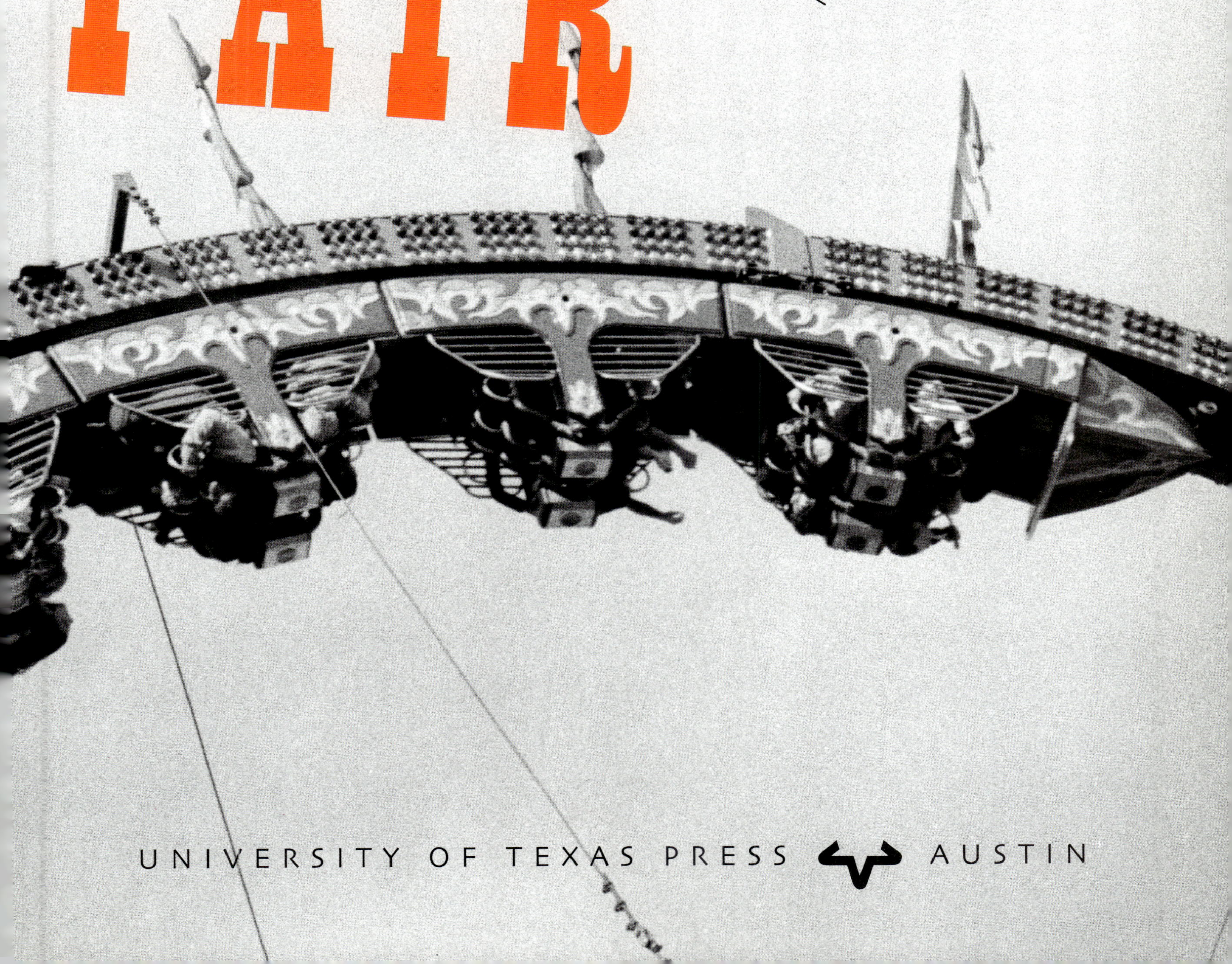

UNIVERSITY OF TEXAS PRESS AUSTIN

Focus on American History Series
Center for American History
University of Texas at Austin
Edited by Don Carleton

University of Texas Press, Austin

Printed in China
First edition, 2006

Requests for permission to reproduce material from this work should be sent to:
Permissions
University of Texas Press
P.O. Box 7819
Austin, TX 78713-7819
www.utexas.edu/utpress/about/bpermission.html

The paper used in this book meets the minimum requirements of ANSI/NISO Z39.48-1992 (R1997) (Permanence of Paper).

Library of Congress Cataloging-in-Publication Data

Grace, Arthur.
State fair / Arthur Grace.— 1st ed.
p. cm. — (Focus on American history series)
ISBN 0-292-71287-1 (cloth : alk. paper)
1. Agricultural exhibitions—United States. I. Title. II. Series.
S554.G73 2006
630'.7473—dc22
2005029157

Designed by Alex Castro, Castro/Arts LLC, Baltimore

STATE FAIR

To Debra,
who helped me find my truth again
and turn it into this book

STATE FAIR

I WAS HOOKED ON STATE FAIRS the minute I walked into the dairy pavilion at the Minnesota State Fair in the summer of 1977. A disembodied voice was booming over the loudspeakers giving a "play-by-play" of the butter sculpture competition. I thought I was hearing things, but as I moved my way through the crowd, sure enough there was an earnest young man having at it with a sculpting blade on a huge block of refrigerated butter that was slowly changing into a bust of a beautiful woman. I didn't know whether to laugh or applaud. I'd never seen anything like it before. Mesmerized, I finally put the camera to my eye to record what was happening, mindful of the old saw "the camera never lies."

I spent the rest of the afternoon wandering the many acres of the fairgrounds, taking in the sights, sounds, and smells of this bizarre and eclectic world. There were 4-H kids sleeping in the barns on stacks of hay next to their prize heifers, marching band majorettes standing proudly with their giant trophies at an awards ceremony, farmers peering over and under the latest John Deere tractors and farm machinery, families sitting on the grass eating corn dogs and cotton candy, people whooping and hollering along the midway as they tossed rings and threw softballs trying to win giant teddy bears. Everywhere you turned there were eager and excited faces, laughter and also looks of exhaustion from too much of a good thing. All I remember as I made my way back to my rental car to head to the airport was how good it felt being there. There was something comforting and reassuring about this state fair business, where people could forget about their troubles, put aside the rigors of everyday living, and come to a place seemingly lost in time where all you were expected to do was simply have fun.

But for a visual person such as myself, there was something more going on, something that I couldn't quite put my finger on at the time. Years later, when I was researching this project, I ran across a quote from a *Washington Post* article by Christopher Corbett on the Ohio State Fair that summed up what I was feeling that day after my first exposure to a state fair. "It is a showplace of the spirit of Middle America . . . celebrating the best of the American heartland in ways that are both endearing and off-the-wall—often at the same time." It was that oddly American juxtaposition of the heartfelt and the huckstering, the totally weird and the comfortably familiar, the love of country and the love of self, the individualism and the mass marketing that made the extravaganza of a state fair so compelling to me and such a mother lode for visual exploration and photographic interpretation.

Over the years as I went on with my career as a magazine photojournalist for *Time* and *Newsweek*, traveling on assignment to points around the globe, I kept thinking about doing a story on state fairs, a quintessentially American phenomenon, but I never found the time. During the national election years of '80, '84, and '88, I parachuted into various midwestern state fairs with presidential and vice presidential candidates as part of the traveling media horde that descends upon unsuspecting events and transforms them for a few hours into spectacles. In the brief time I spent at these fairgrounds, I was once again captivated by the action, diversity, smells, and sounds of these uniquely American celebrations. Finally, in 2003, I had the time and opportunity to head out on the road and explore what had been intriguing me for so many years and had made me smile any time I'd heard or read the words "state fair." What was so compelling about these annual gatherings that they pull in more than twenty million Americans every year? Why were these convocations such a magnet for so many people in every geographic area of the country?

The first thing academics will tell you is that going to the state fair is a ritual, although the people I talked to at the fairs kept using the term "tradition"—"We always go every year" . . . "I've been going since I was a little kid" . . . "It's a tradition in my family." It seems that if your parents took you on an annual pilgrimage to the state fair when you were growing up, you're likely to bring your own kids when you become a parent. This wasn't the case with me, growing up in Worcester, Massachusetts. The closest I came to anything resembling a state fair was a third-rate circus that pulled into town once a year with a trainload of over-the-hill jungle animals too tired to growl or roar. (My only other experience was an evening visit to Maine's Topsham County Fair with some inebriated college fraternity brothers to do field research on "exotic dancers.") The more I talked to people, the more I heard the idea of the state fair being a "tradition": "I come back every year for the corn dogs" . . . "I love to watch the pie judging" . . . "I like to see what's new from John Deere" . . . "We can't wait to get back to the midway and go on the rides" . . . "I haven't missed a Grand Champion hog competition in eighteen years."

There's no question, however, that the underlying reason why people flock to state fairs is the pursuit of that time-honored, all-American combo of FUN & FOOD (not necessarily in that order). People intuitively realize that when they enter the fairgrounds they have only two obligations: to have a great time and to eat till they drop. The only anxiety you sense among new arrivals at a state fair is the confusion that's caused by the sudden sensory overload: hucksters on loudspeakers, the thunder of the midway rides, the crying, laughing, and screaming of kids, the blaring trumpets of high school marching bands, the eclectic musical overload of myriad singers and dancing troupes, and of course the hundreds of signs and flashing lights urging you to come see this or eat that. It's a daunting task coping with so much to see and do over so many acres, while at the same time dealing with a constant barrage of seductive aromas from all the standard "fair food"—corn dogs, ice cream, gyros, cotton candy, cheese fries, smoked turkey legs, pizza, barbecued ribs, popcorn, fried chicken, strawberry shortcake, deep-fried Twinkies, Polish sausage, roasted corn on the cob, and saltwater taffy, among other delectables.

Not to worry. In short order, intrepid fairgoers manage to rise to the challenge and begin eating their way through the fair while enjoying all manner of entertainment laid out in front of them. There's no heavy lifting required. Just pick your pleasure or go for whatever piques your interest. Scattered along the paths and streets and roadways are a never-ending series of sideshows featuring jugglers and acrobats, clowns and escape artists, magicians and alligator wrestlers, bear trainers and bird wranglers. Every day there's a major parade with marching bands, cheerleaders, a beauty queen or two, and plenty of American flags. On the dancing front alone, I saw performances by line dancers, tap dancers, clog dancers, square dancers, break dancers, ballroom dancers, flamenco dancers, polka dancers, and folk dancers. I heard all kinds of music, from karaoke to rap to yodeling to rock to country to jazz to bluegrass to church choirs—all performed live at pavilions and stages and halls throughout the fairgrounds. And the good news? It was all free (although on occasion it became painfully clear why local talent stays local).

For those who need their fun balanced by some cerebral stimulation or who have a twinge of guilt at their self-indulgence, they can take comfort in the mantra that state fairs are "educational" (and are often billed that way). There is much to be learned at a state fair if you're so inclined. Without even trying, just by walking around the fairgrounds taking photographs, I couldn't help but soak up all kinds of knowledge. There was the 4-H pavilion, where I learned about rabbit care and anatomy during "show-and-tell" competitions; the "Bug World" house, where I learned about the habits of African centipedes and American cockroaches; the "Honey Booth," where I learned about the wiles of the queen bee and how not to incite her drones; the "Food Pavilion," where I learned it was possible to make a variety of wonderful meals by using only Del Monte canned goods. On the medical front I learned about the warning signs of a heart attack, how to recognize skin cancer, and how to take my blood pressure at home. As a dog lover, I got tips on obedience training (for myself as well as my English bulldog, Cleopatra) and had the once-in-a-lifetime experience of watching a live spaying operation on a large alley cat, performed by a "special guest" veterinarian. And I heard over and over again from a twenty-foot-tall Smokey the Bear with a voice loop about how I could prevent forest fires.

In the great American tradition of free enterprise and unbridled capitalism, state fairs are also sales fairs. For anyone coming through the gates with some cash burning a hole in their pocket who can't wait to buy the latest gadget, or cure, or "revolutionary breakthrough"—welcome to the promised land. Entire buildings and huge tents are chockablock with salespeople touting everything from automatic massaging tables, to super mops, to eyeglass defoggers, to telescoping flagpoles, to hand-carved life-sized replicas of horny toads. You're limited only by how much you can carry and how quickly you can max out your credit cards. I've never seen so many people spellbound by the hypnotic cadences of world-class sales pitches hyping everything from "miracle" sponges that can soak up a pail of water in one go to a "diet belt" that can melt off the pounds while you're sleeping. Remarkably, nobody snickers or laughs at the end of these spiels, and the only sound you hear after the applause has died down is the constant clack-clack of the well-greased credit card machines sliding back and forth running up sales. (I myself succumbed to an Australian wide-brimmed bush hat that can fold up to the size of a napkin.)

The more days I spent wandering around state fairs all over the country in my official role as a professional observer (and an accredited member of the media), the more I came to realize that the real backbone of state fairs, and what makes them unmistakably American, is their celebration of "competition"—any and all kinds, highbrow and lowbrow, from the mundane to the inane. Americans don't care what the contest is so long as they are either competing in it or getting to watch someone else compete. Even though rooting for the underdog is ingrained, Americans love a winner. This striving to be the best, to be #1, is embedded in the DNA of the great American gene pool, and the state fair is definitely a showcase for it.

Naturally, state fairs have all the macho, rough-and-tumble stuff you'd expect: wrestling, boxing, mutton bustin', and tractor pulling—and that's just the competition for kids under twelve. For adults, there's weight lifting, arm wrestling, demolition derbies, steer roping, and bull riding. On the less strenuous side, you have everything from grocery-bagging competitions (with a chance to go to the national championships in Las Vegas) to 4-H public-speaking contests to horseshoe tossing. And then there are the perennial favorites like corn-dog-eating and pie-baking contests, animal costume and quilt-making competitions, and myriad championships for marching bands, baton twirlers, cheerleaders, and dancers. However, at the heart of every state fair are the dozens of animal competitions—rabbits and goats, poultry and pigeons, sheep and hogs, cows and steers, llamas and geese, dogs and cats, mini-donkeys and horses. Every competitor in this arena, regardless of how old or young, dreams of going home with a Best in Show trophy or a Grand Champion ribbon presented by one of the many specialized fair "queens" (Miss Holstein, for instance).

In the end, when I took a step back from all the commotion and crowds, what struck me the most was how a state fair reflects the American spirit—patriotic, entrepreneurial, fun-loving, competitive, traditional. But above all, a state fair is "democratic"—everyone is welcome and it's affordable for all. Rich or poor, young or old, regardless of color or ethnic background, everyone has the same opportunity to enjoy, to learn, to cheer, to laugh, and to be surprised. I couldn't help thinking that if aliens ever landed on Earth and you only had a few hours to help them understand America, all you would have to do is take them to any state fair. For me, a state fair is a microcosm of America—in all its glory and weirdness—at any given point in time.

First and foremost I'd like to thank Dr. Don Carleton, director of the Center for American History at the University of Texas at Austin. This project would never have seen the printed page without his unfailing support and stewardship.

My thanks to everyone at the University of Texas Press for their skill, diligence, and professionalism throughout the production process.

I'd also like to thank my lawyer and book agent, Paul Mahon, for all his special efforts on behalf of this book and for his unerring advice and good humor. Special thanks to Alex Castro, who once again applied his prodigious talents and unique aesthetic to both the selection of photographs and the design of this book. Thanks also to Jane Livingston and Jo Tartt for taking the time and doing such a superb (albeit ruthless) job of editing the photographs during one very long autumn day in Washington, D.C., and to the multitalented photographer Bill Pierce for his expert advice in all matters photographic and for all his special work in scanning the images and producing digital prints.

I am also grateful for the assistance and support provided by Alain Labbe, owner of Paris Photo Lab in Los Angeles, and Nassi Ashkan, my service representative; Ken Hansen of Ken Hansen New York, Inc.; and Roger Horn, president of Leica Camera, Inc.

And finally, I'd like to thank my sister Dianne Solomon and niece Cara Solomon for their considered "opinions" at key stages in the production of this book; thanks to Ben and Cleo for their canine inspiration; and most of all, thanks to my wife, Debra, for her unswerving support and belief in this project and for her insightful comments as a "first responder" to the raw images each time I returned from the field.

STATE FAIR

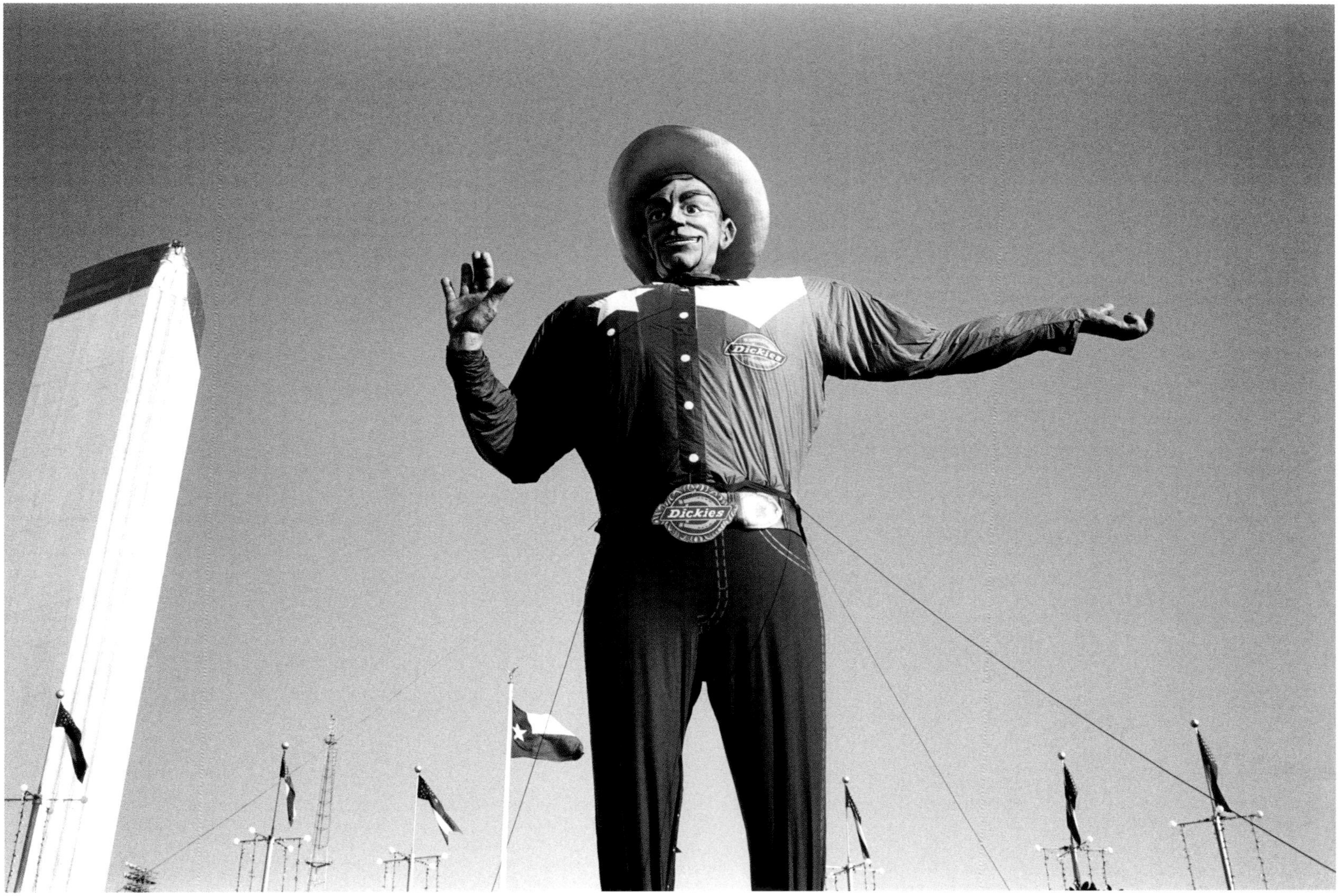
Dickies
Dickies

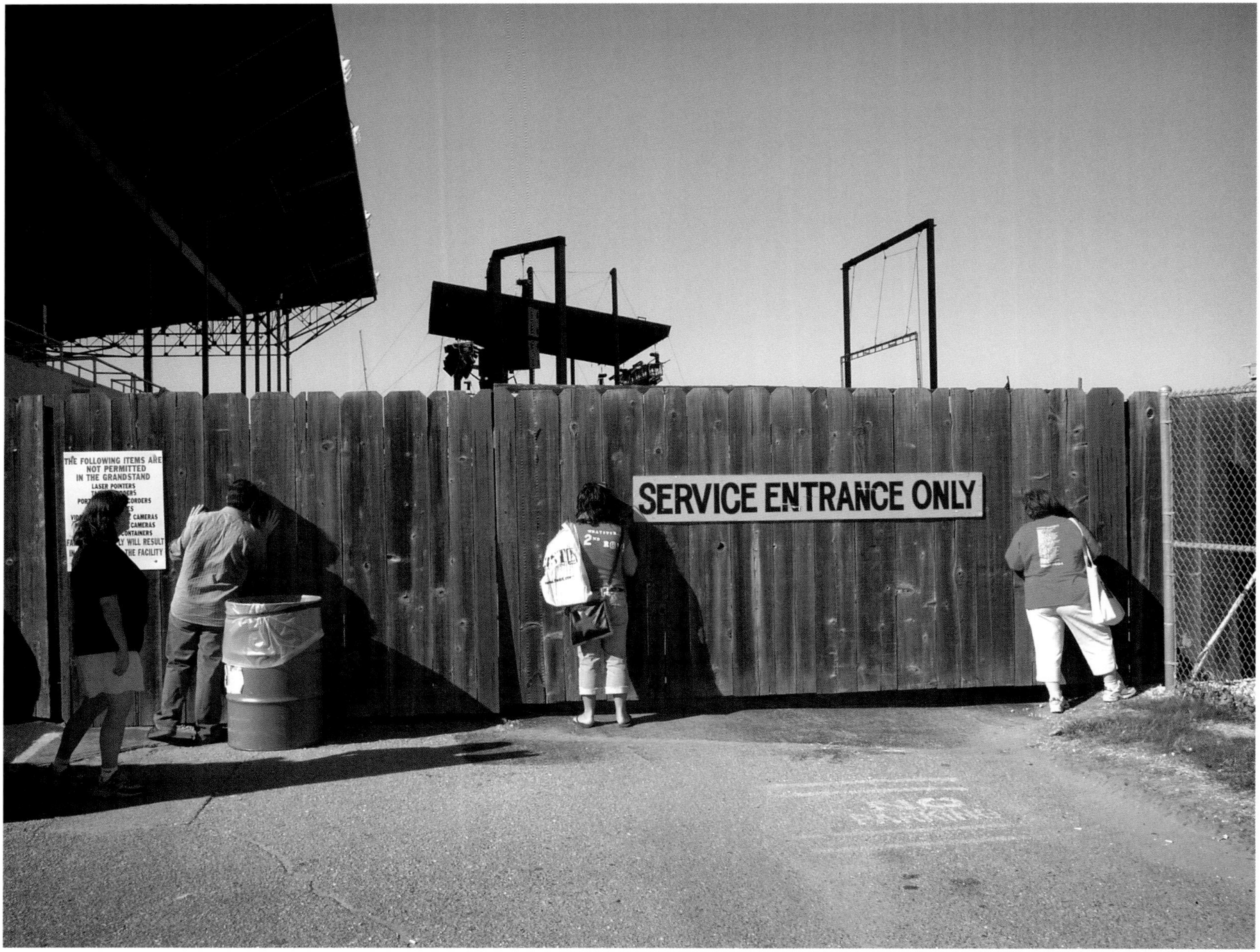
THE FOLLOWING ITEMS ARE
NOT PERMITTED
IN THE GRANDSTAND
LASER POINTERS
SERVICE ENTRANCE ONLY

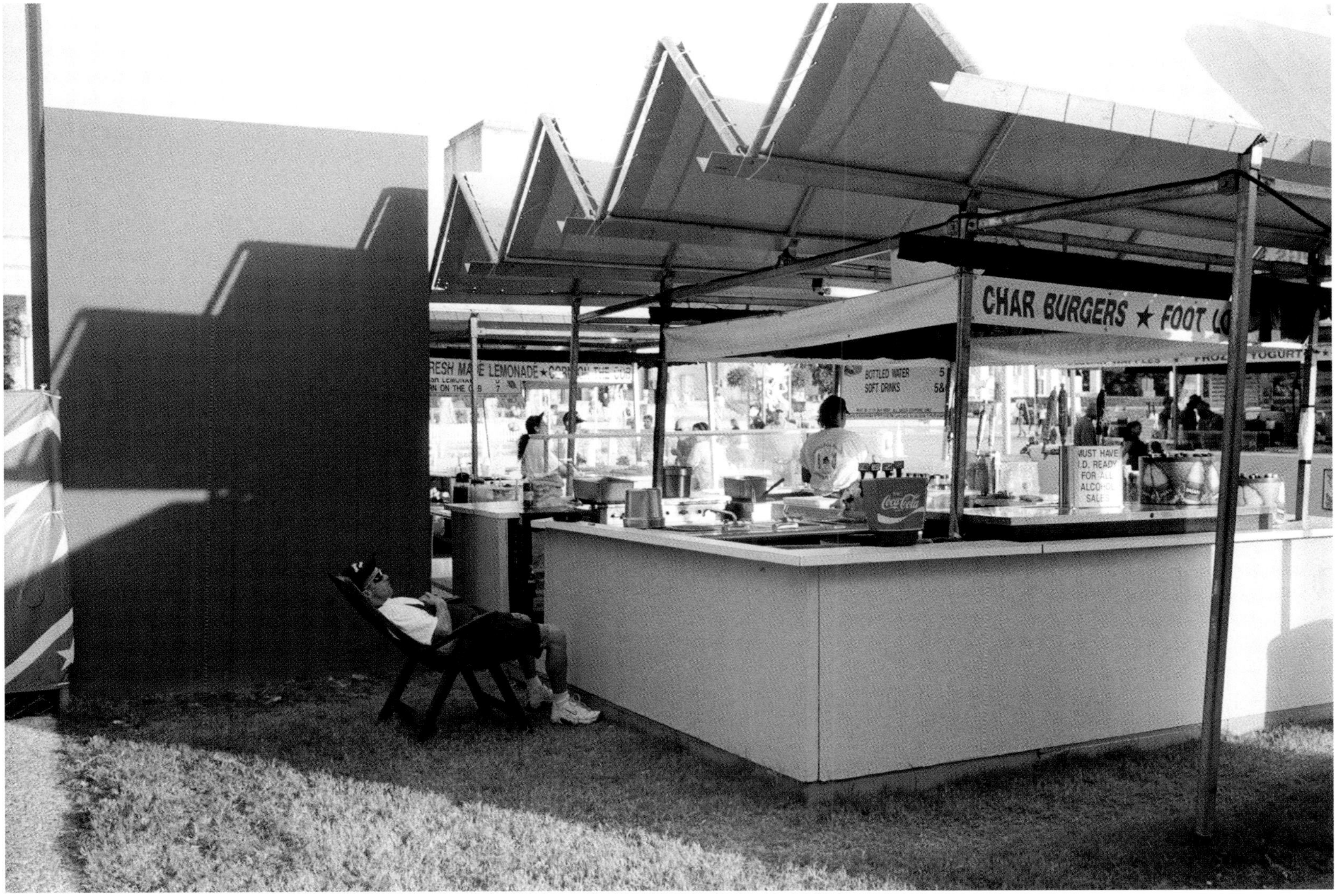
CHAR BURGERS ★ FOOT
BOTTLED WATER
SOFT DRINKS
Coca-Cola

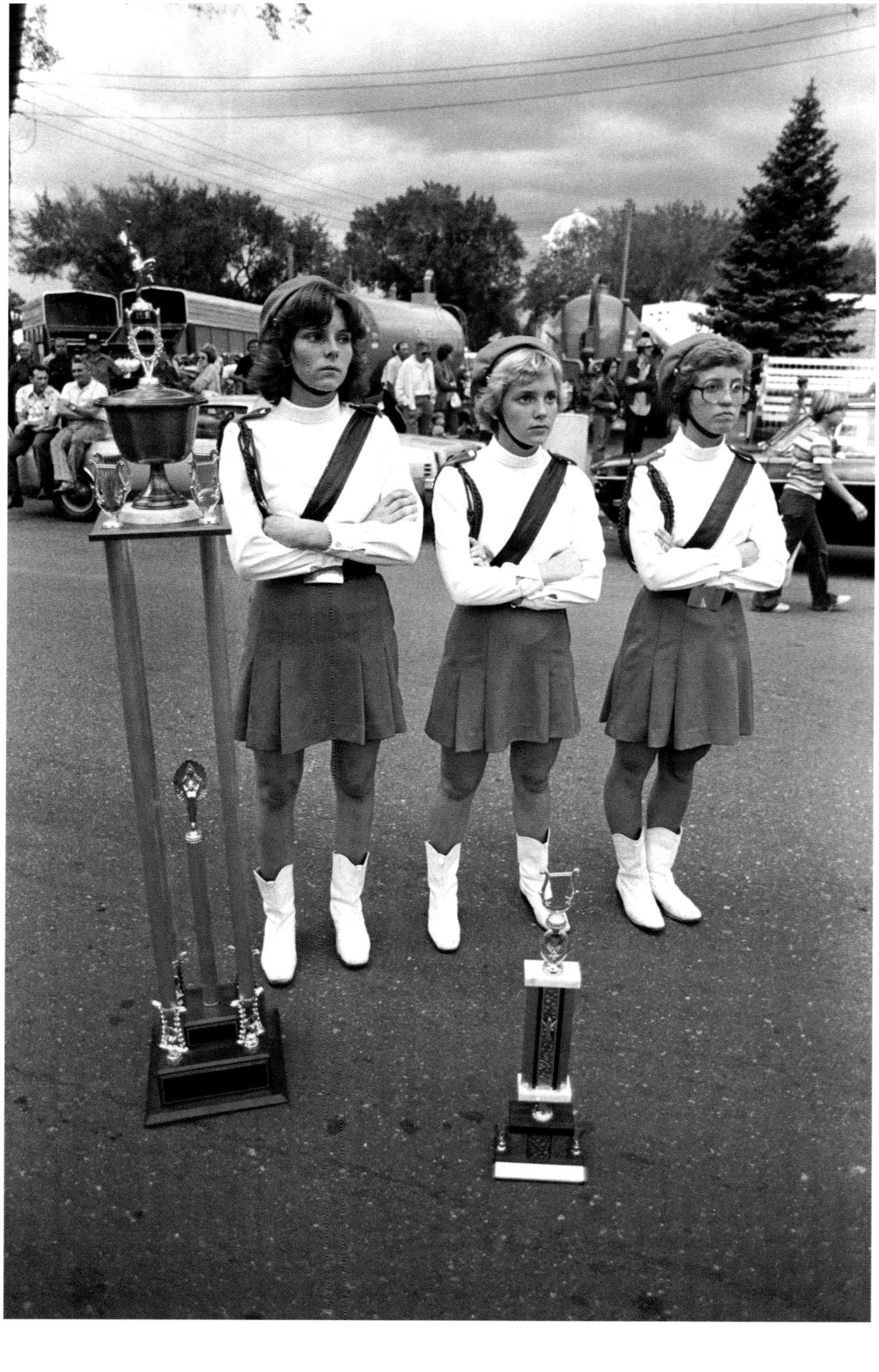

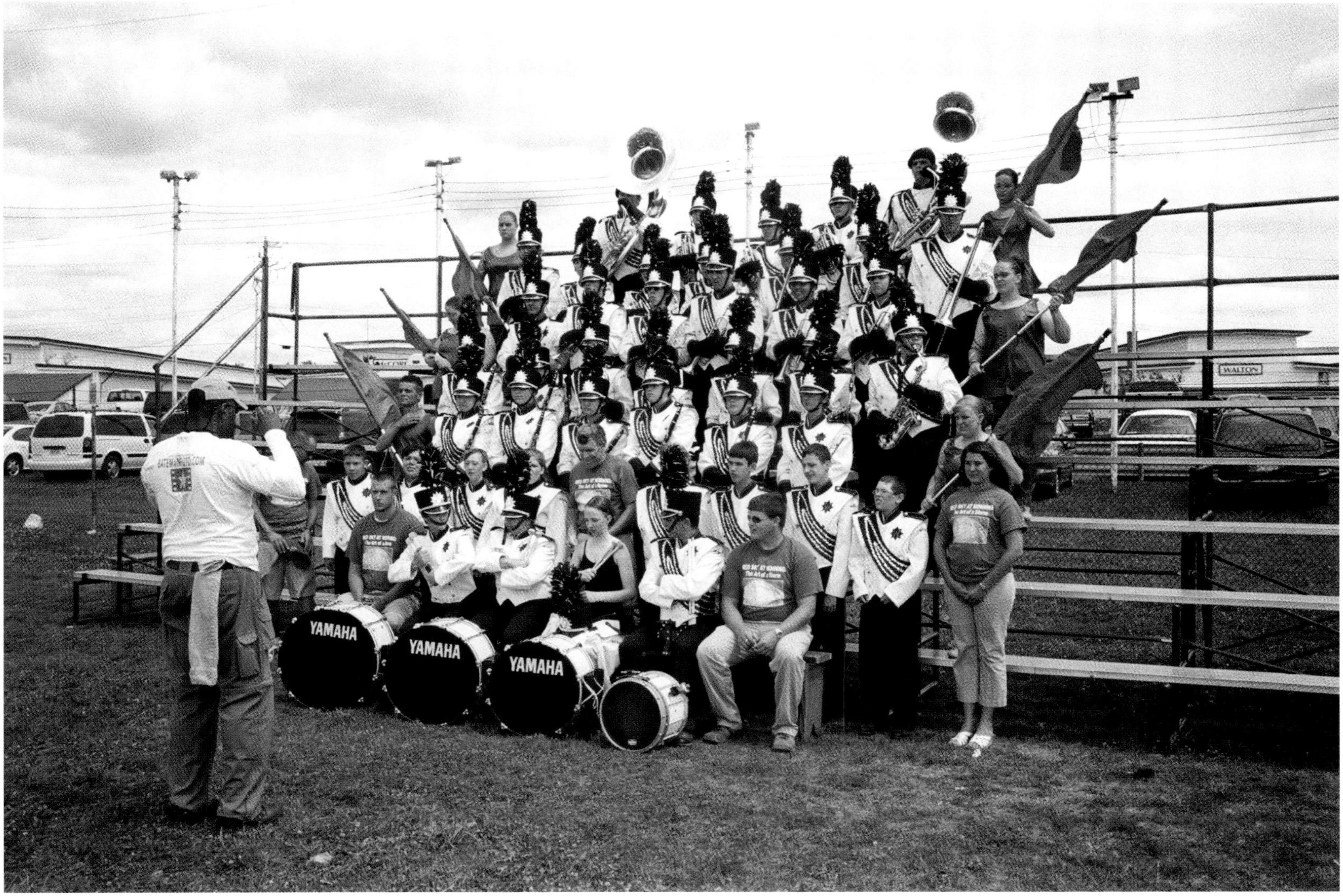
YAMAHA
YAMAHA
YAMAHA
WALTON

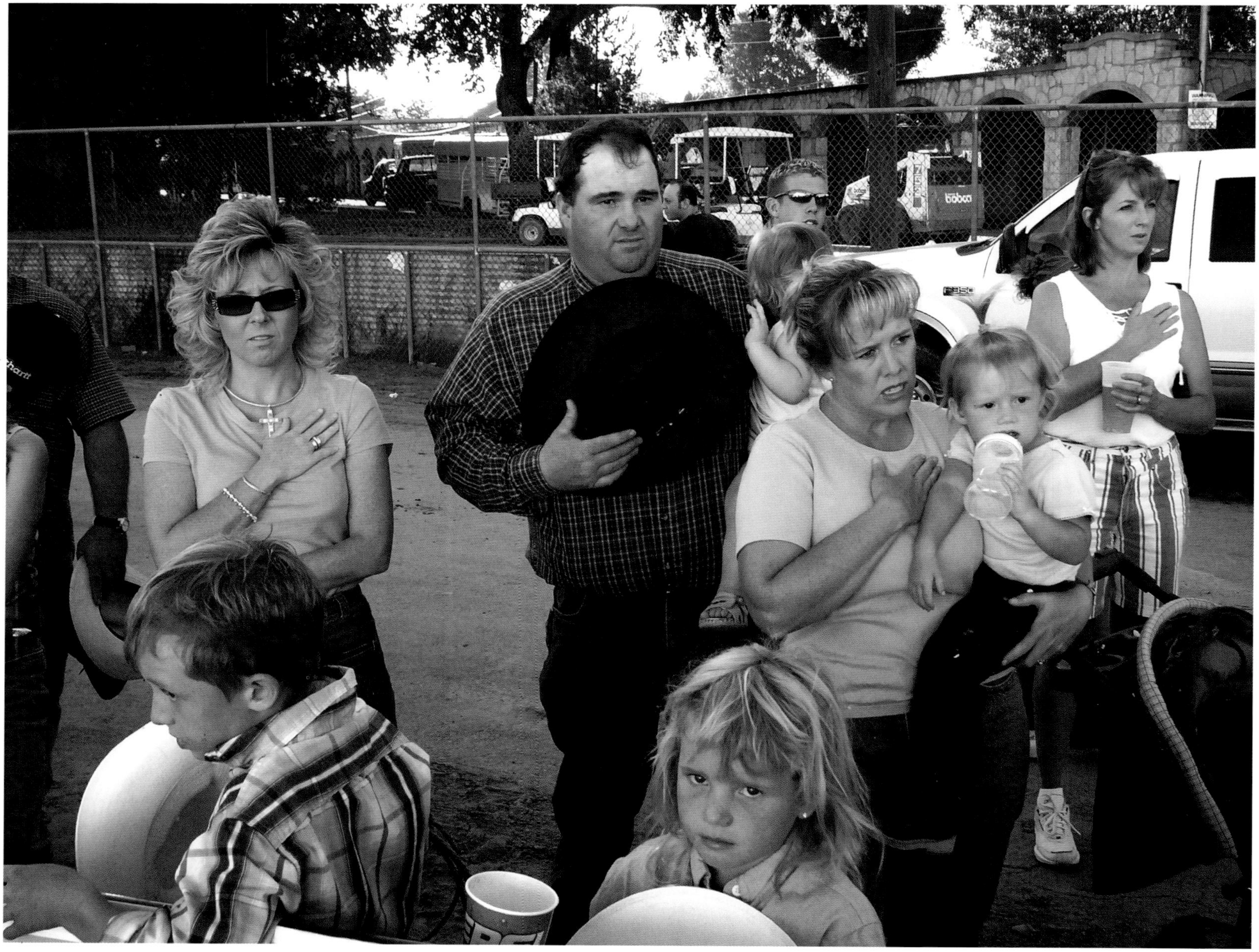

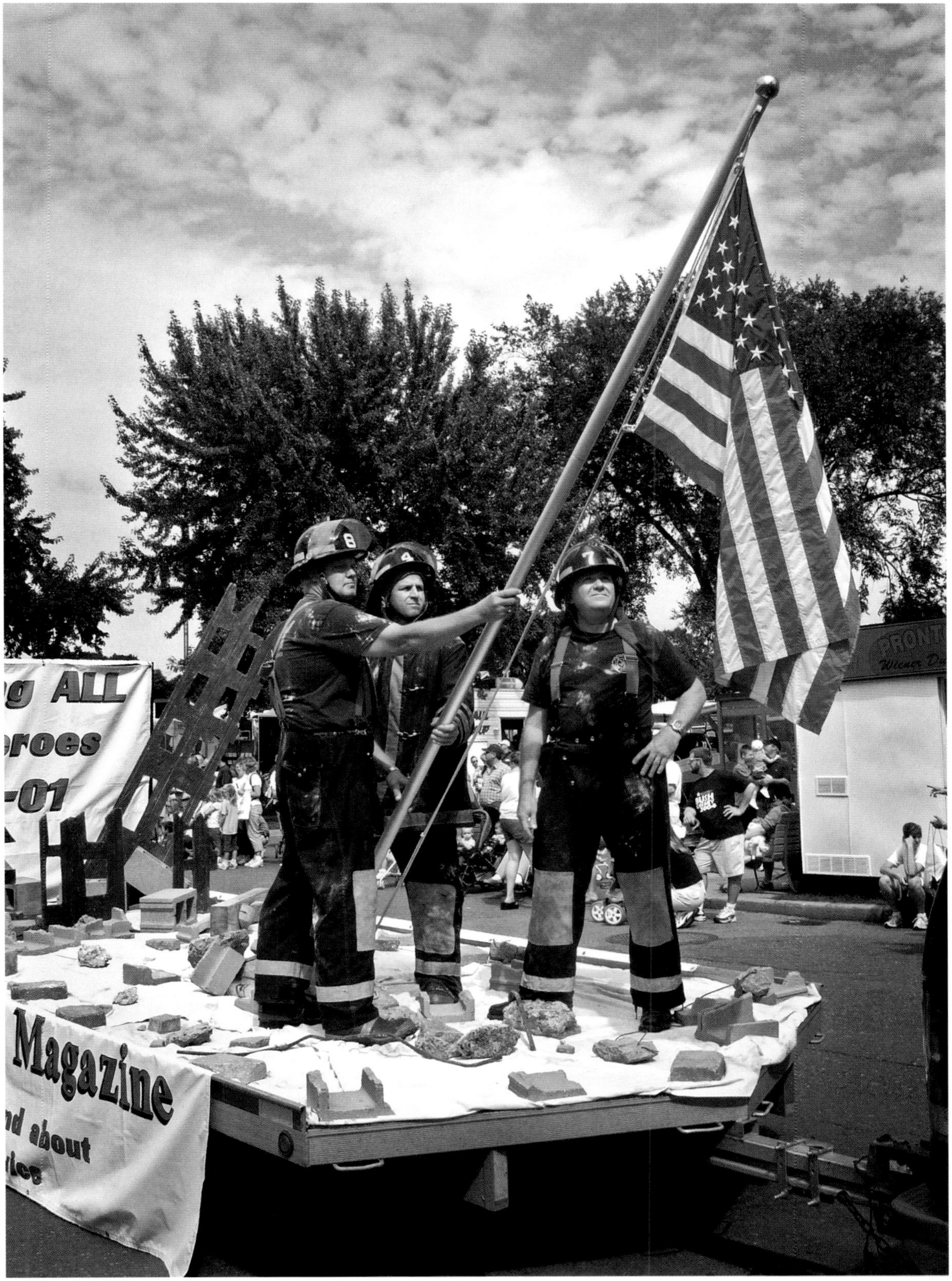
g ALL
eroes
-01
Magazine
nd about

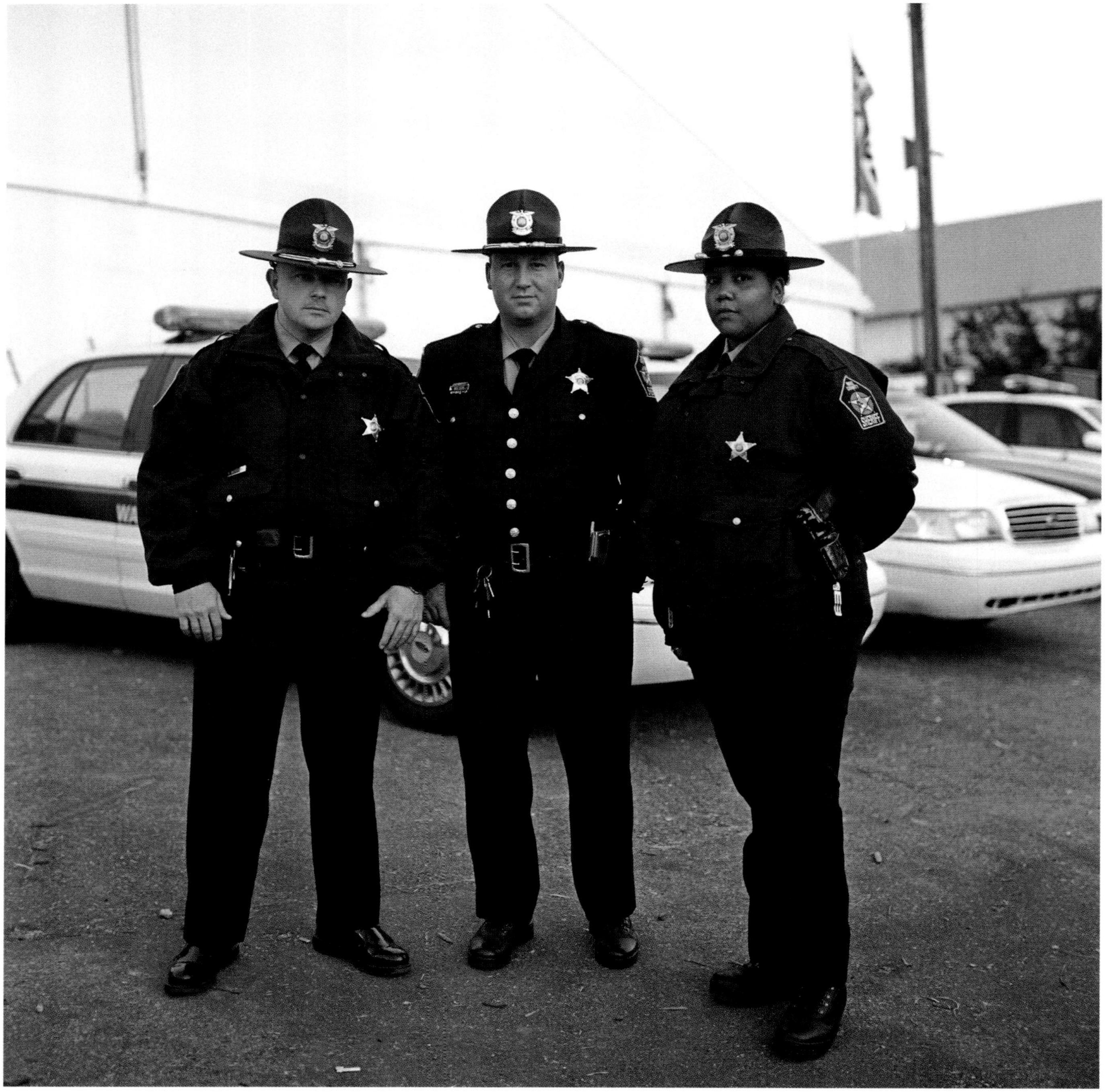

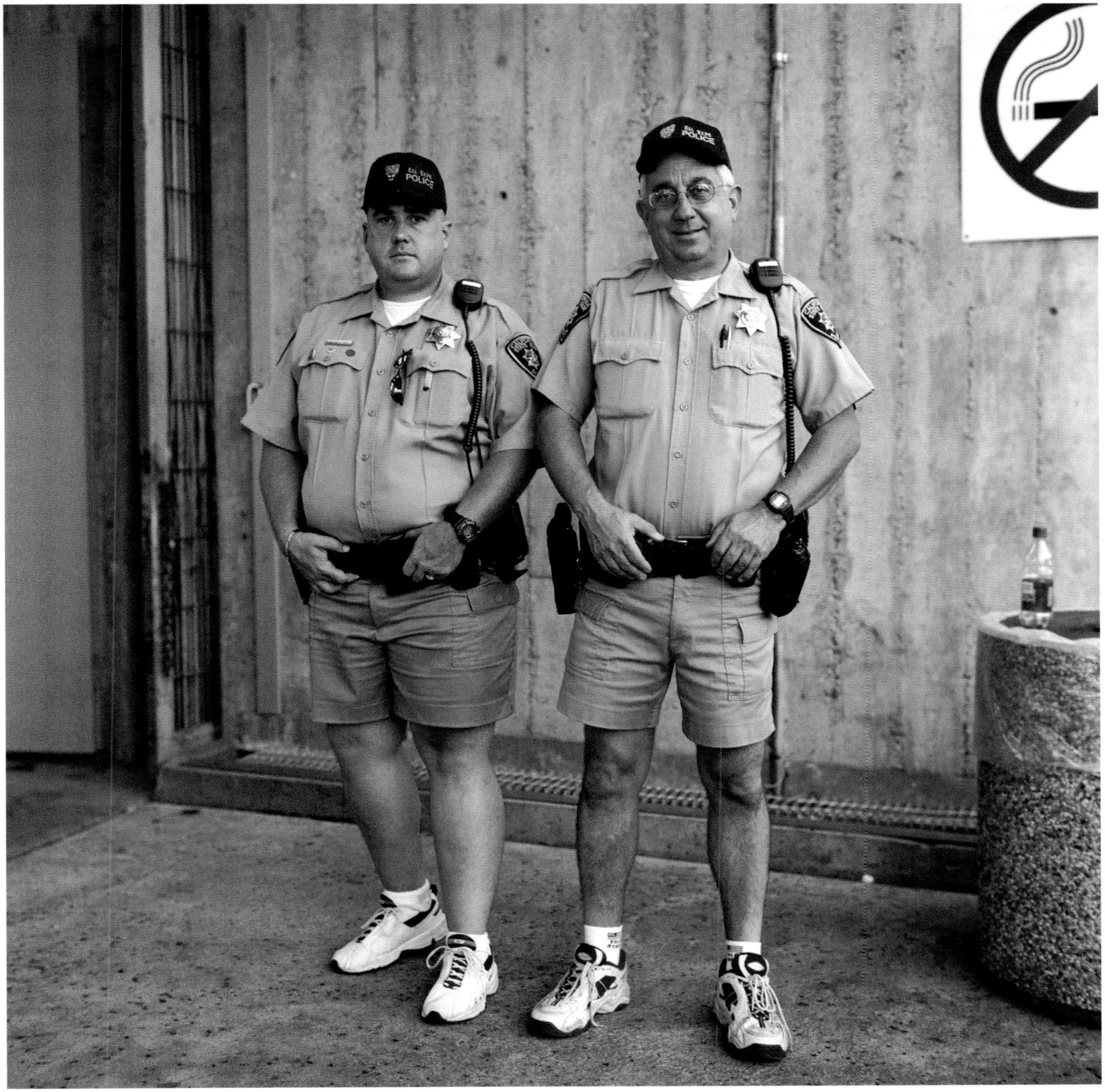
POLICE
POLICE

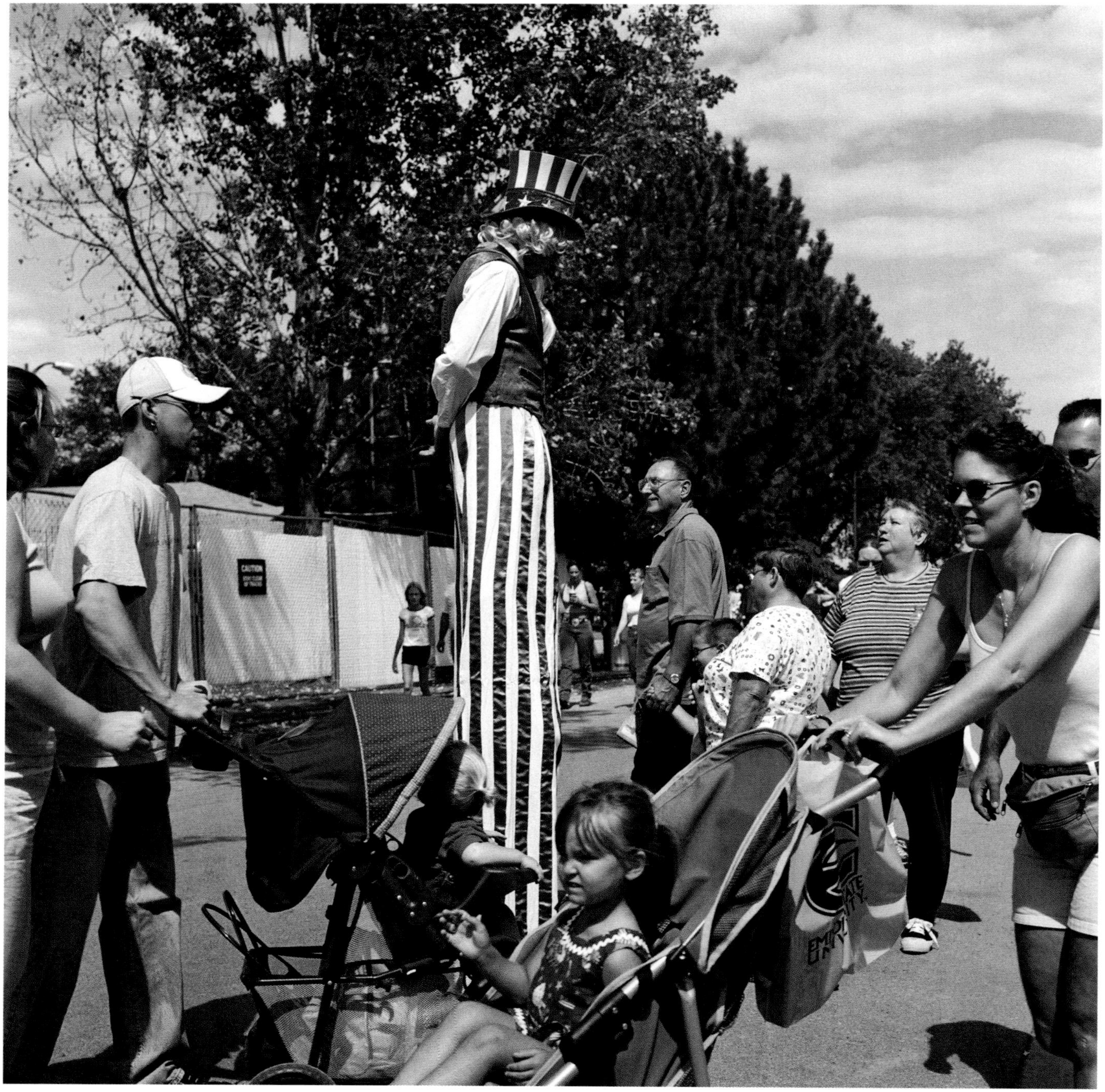

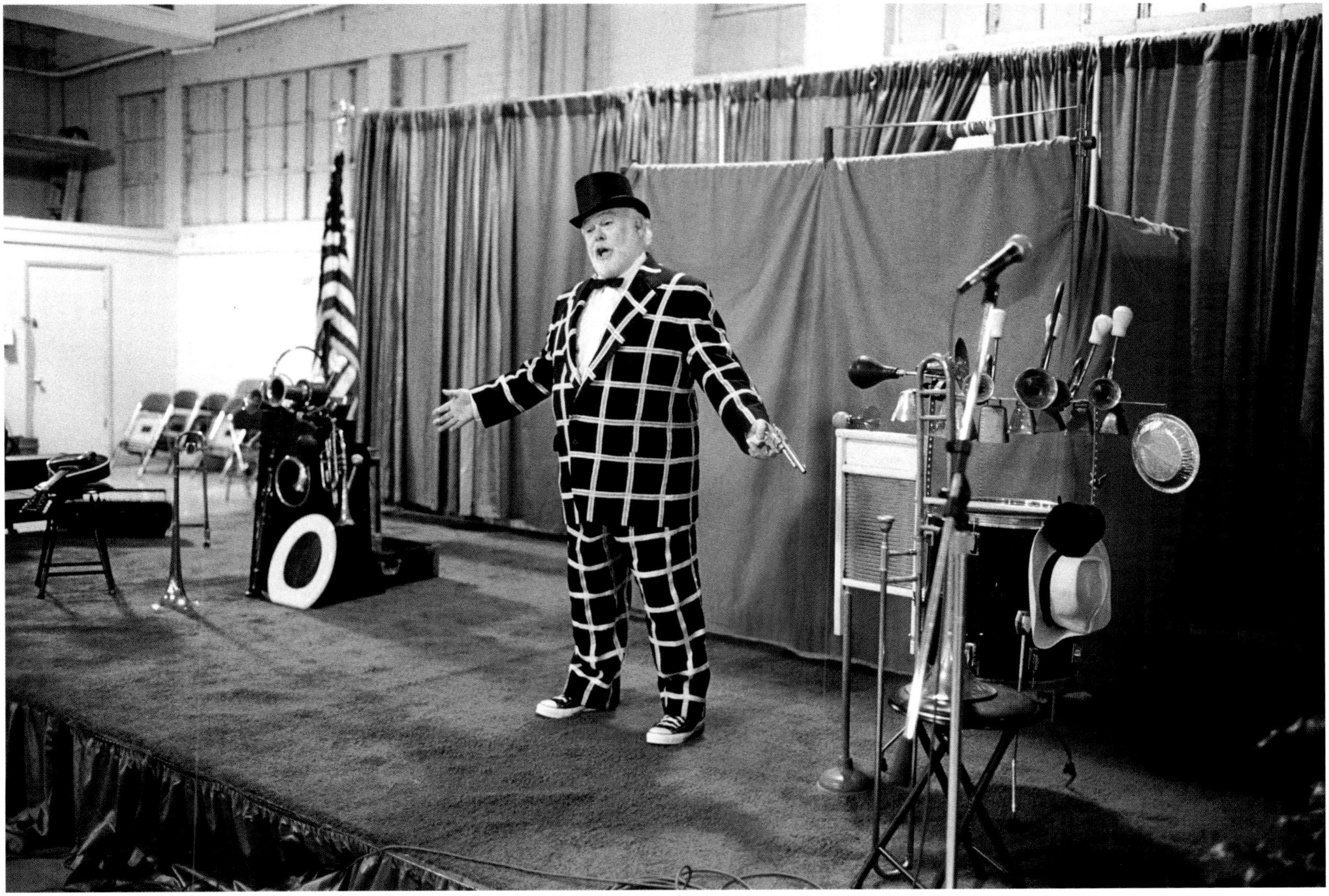

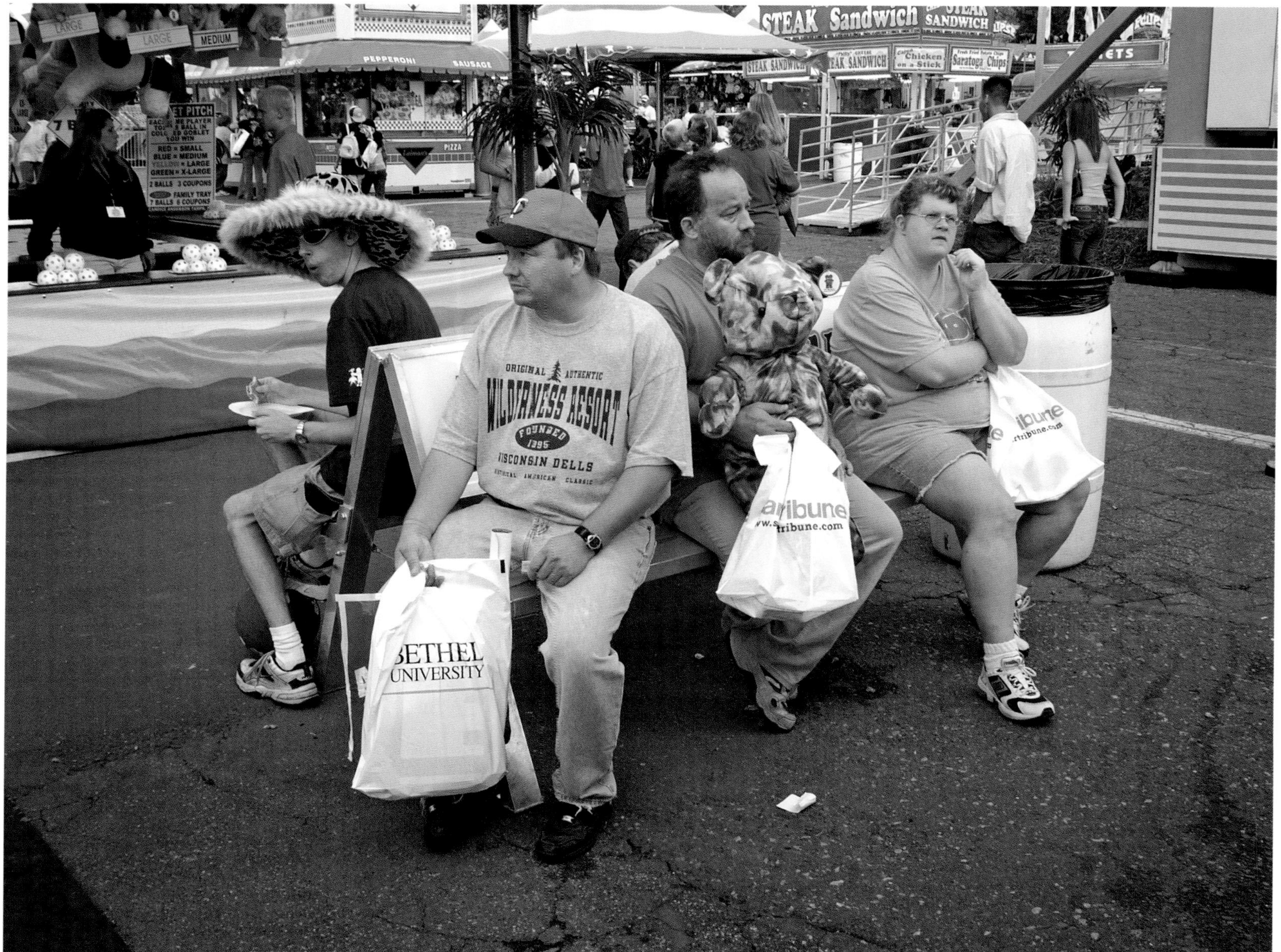
LARGE
MEDIUM
PEPPERONI
SAUSAGE
PIZZA
RED = SMALL
BLUE = MEDIUM
GREEN = X-LARGE
2 BALLS 3 COUPONS
FAMILY TRAY
7 BALLS 6 COUPONS
STEAK Sandwich
STEAK SANDWICH
STEAK SANDWICH
STEAK SANDWICH
Chicken on a Stick
Saratoga Chips
ORIGINAL AUTHENTIC
WILDERNESS RESORT
FOUNDED 1995
WISCONSIN DELLS
BETHEL UNIVERSITY

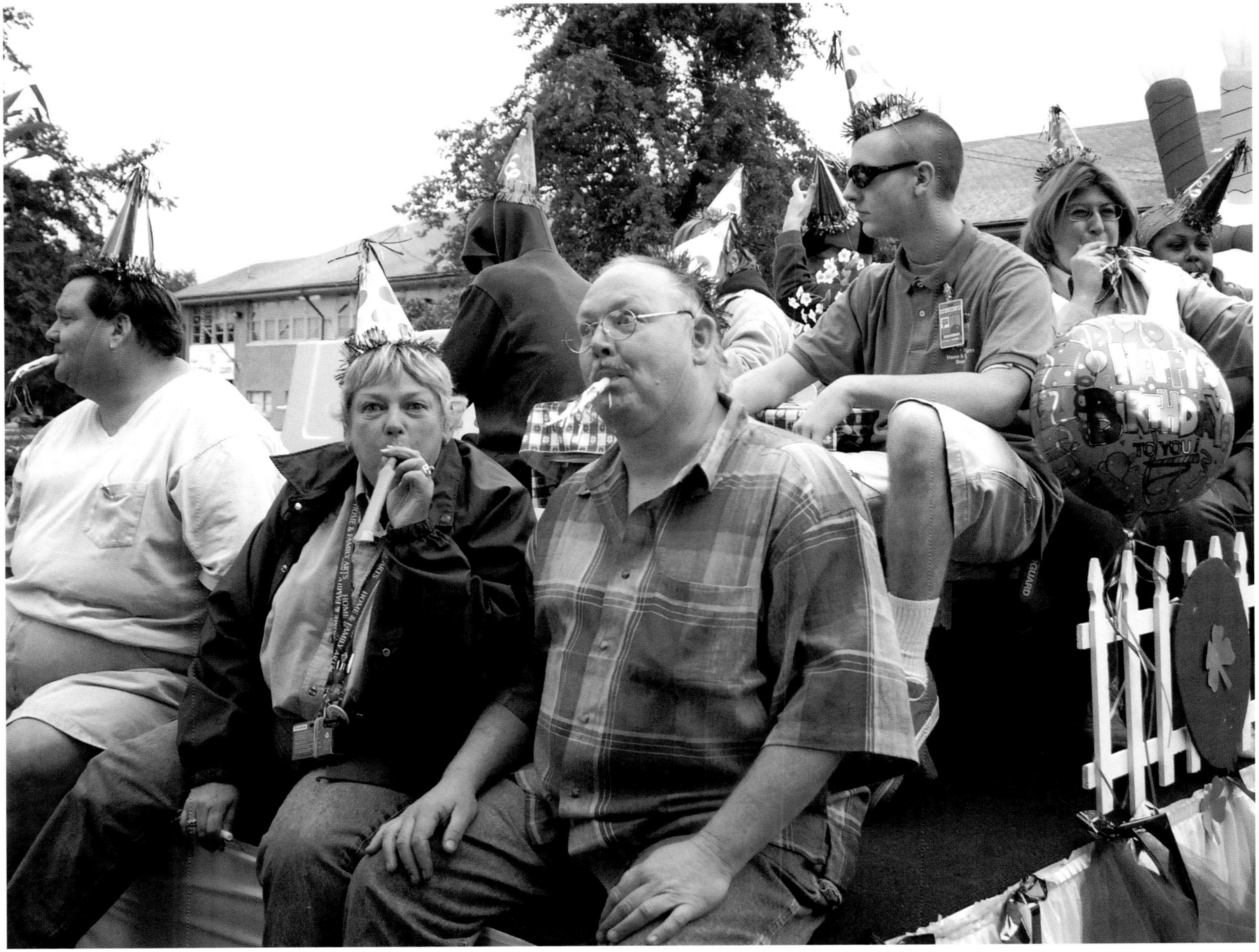
TO YOU!

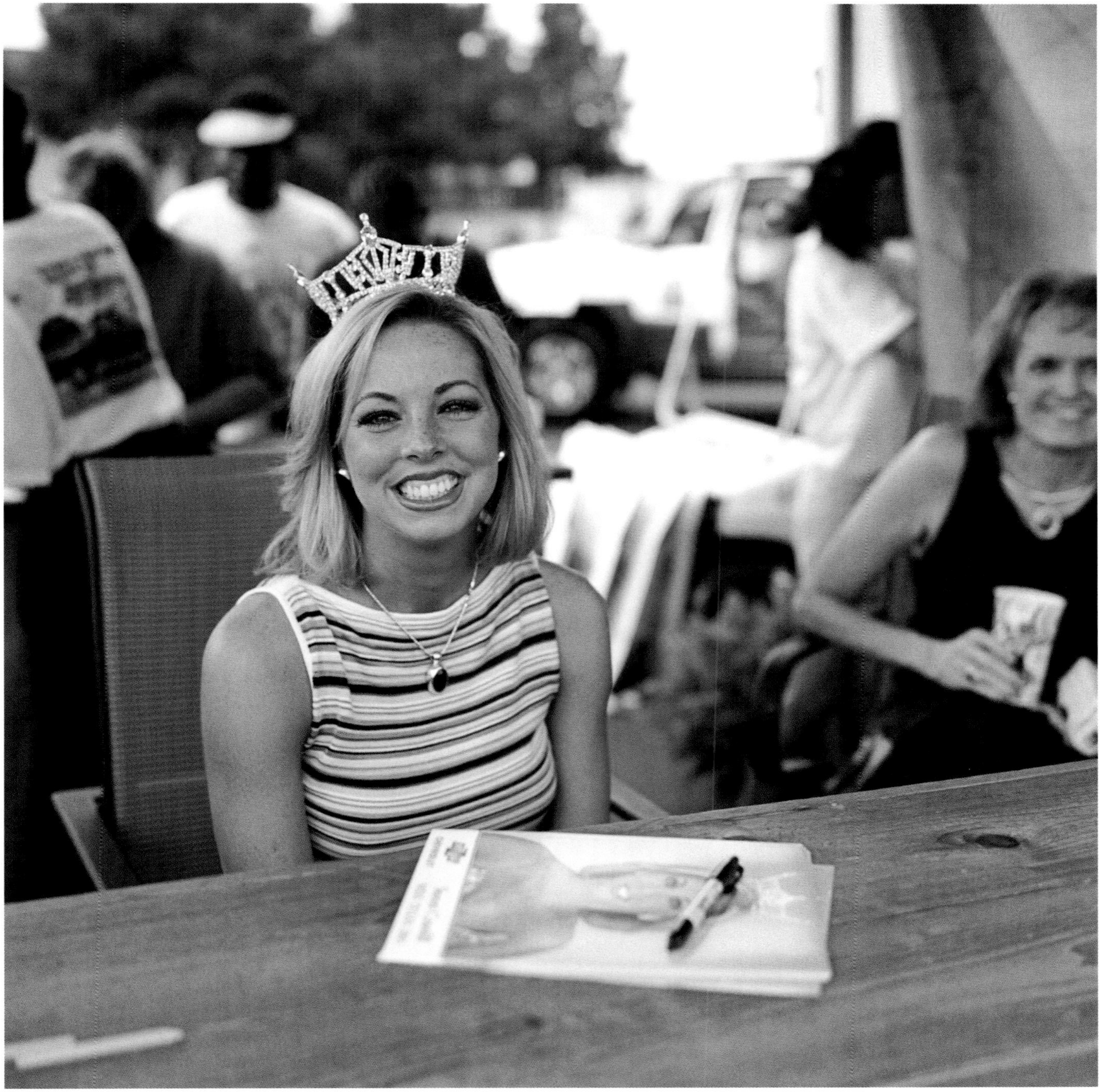

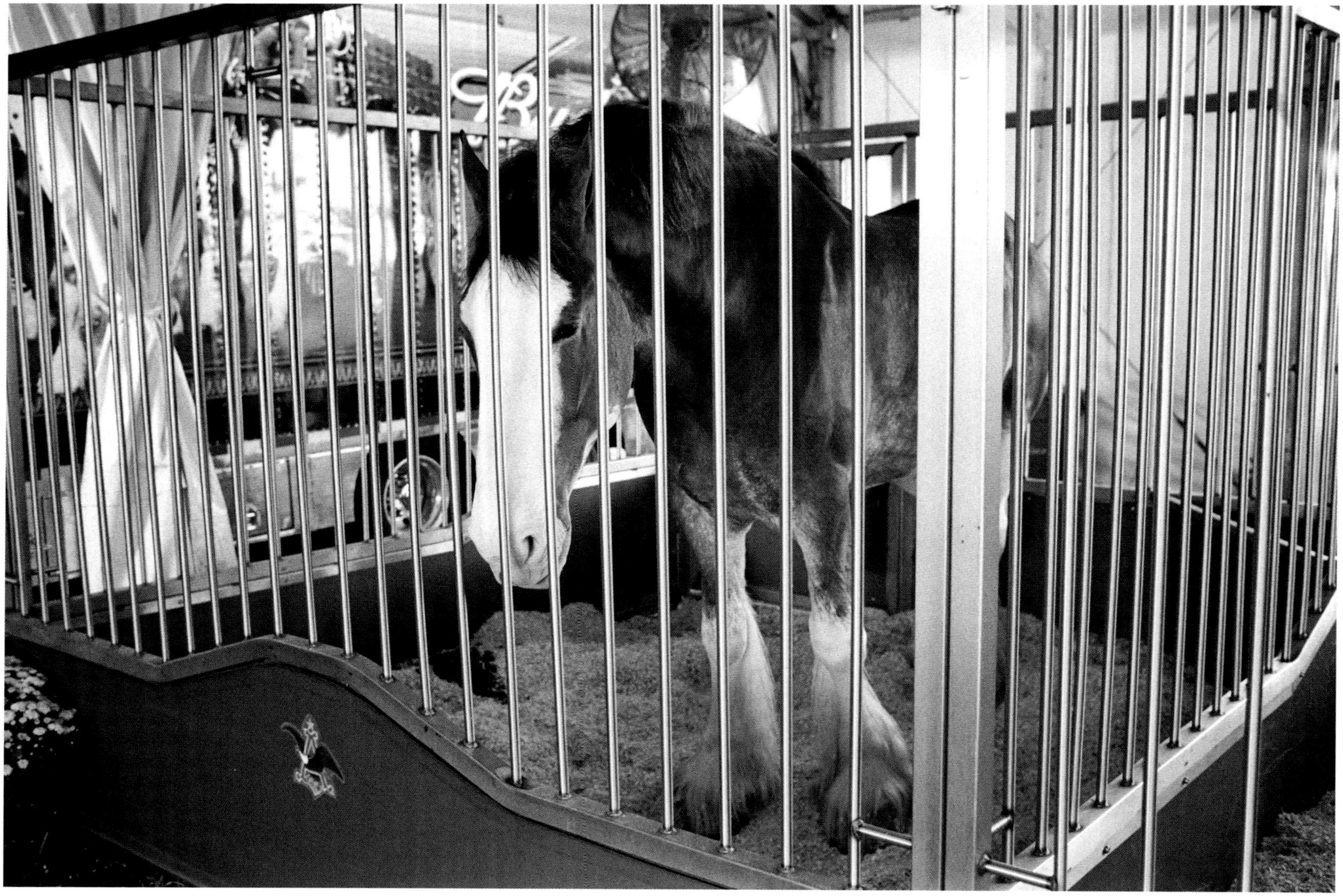

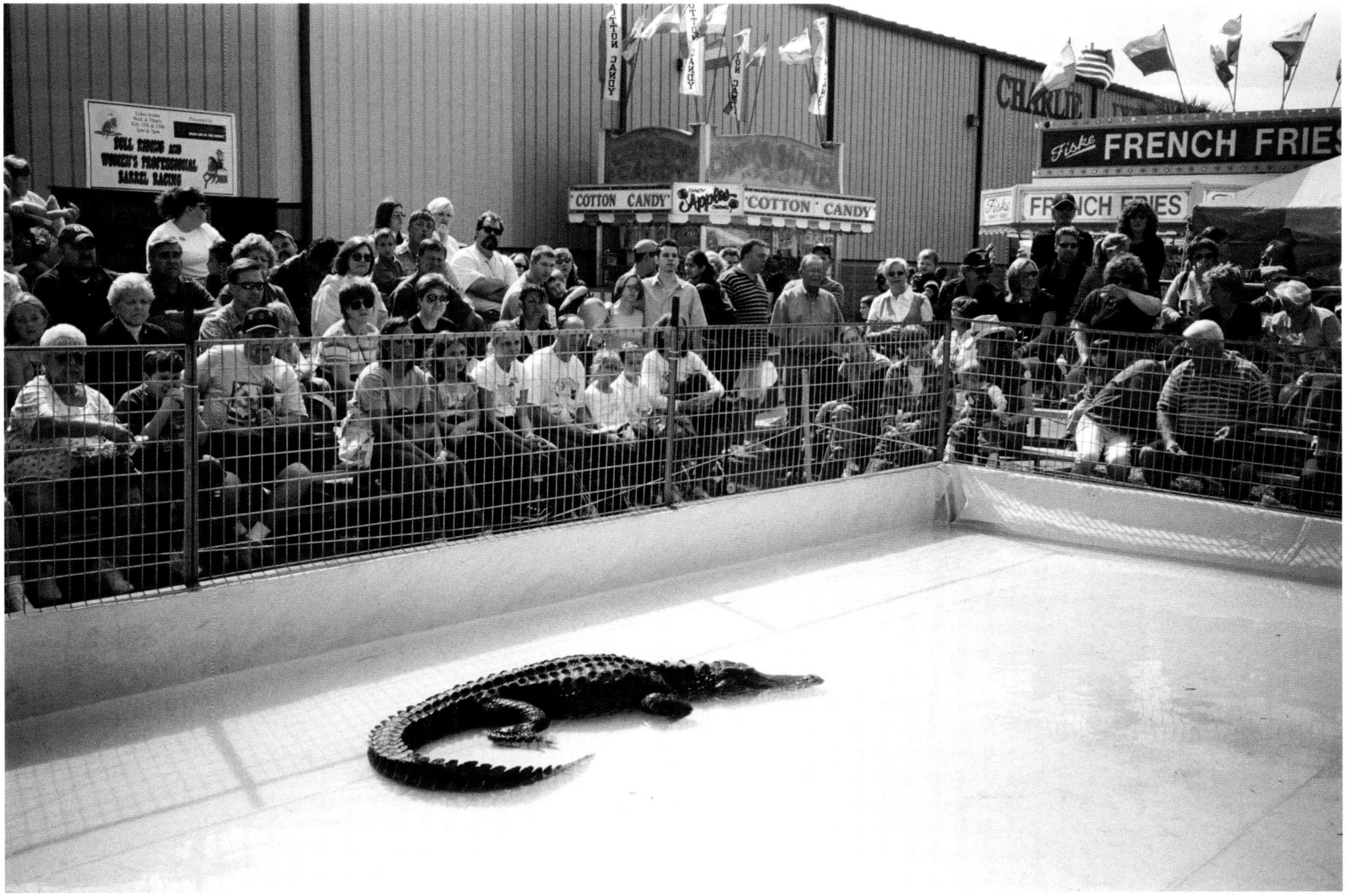
Fiske FRENCH FRIES
COTTON CANDY
COTTON CANDY
FRENCH FRIES

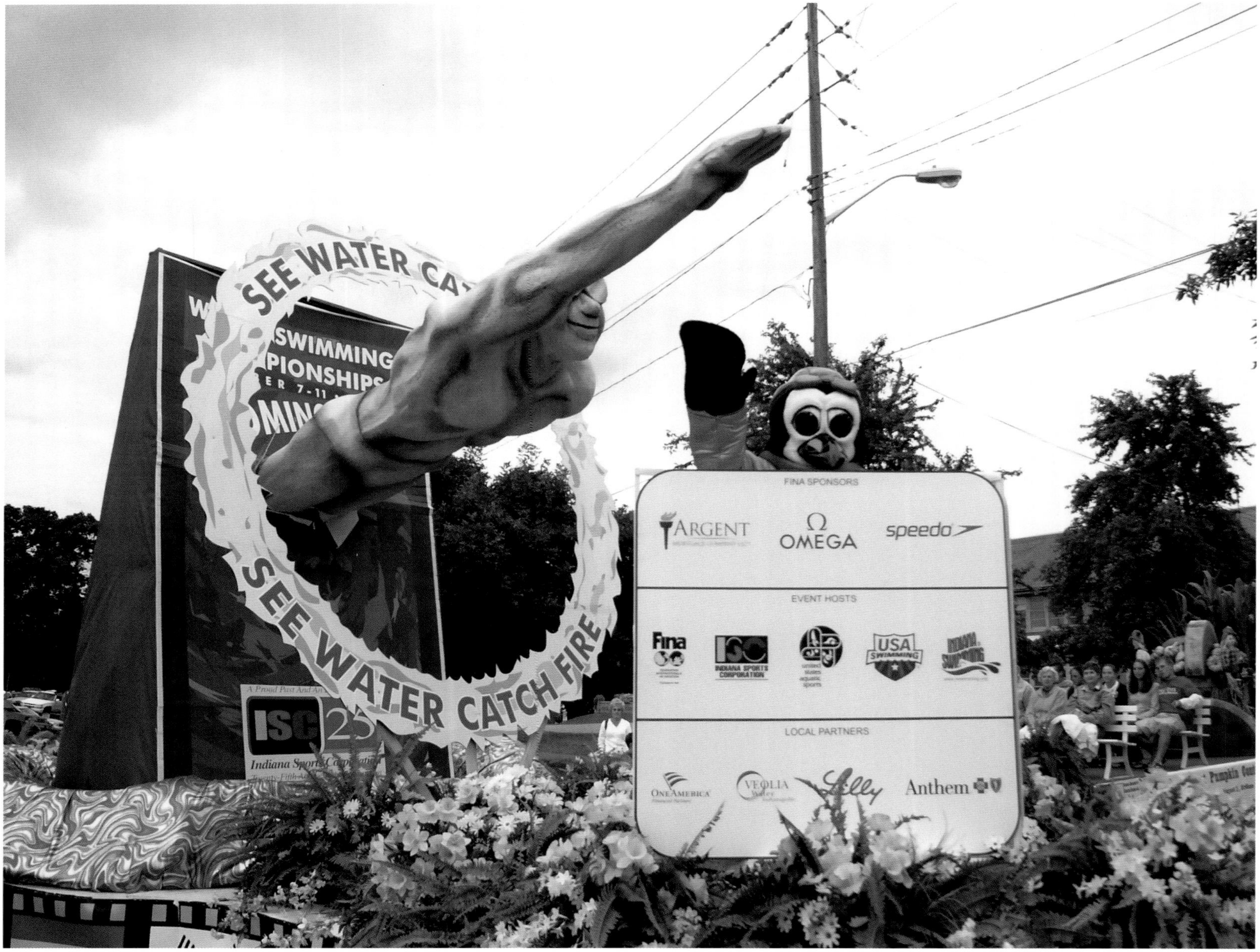
SEE WATER CA
SWIMMING
PIONSHIPS
SEE WATER CATCH FIRE
A Proud Past And An
ISC 25
Indiana Sports Corporation
FINA SPONSORS
ARGENT
OMEGA
speedo
EVENT HOSTS
Fina
INDIANA SPORTS CORPORATION
USA SWIMMING
LOCAL PARTNERS
OneAmerica
VEOLIA
Anthem

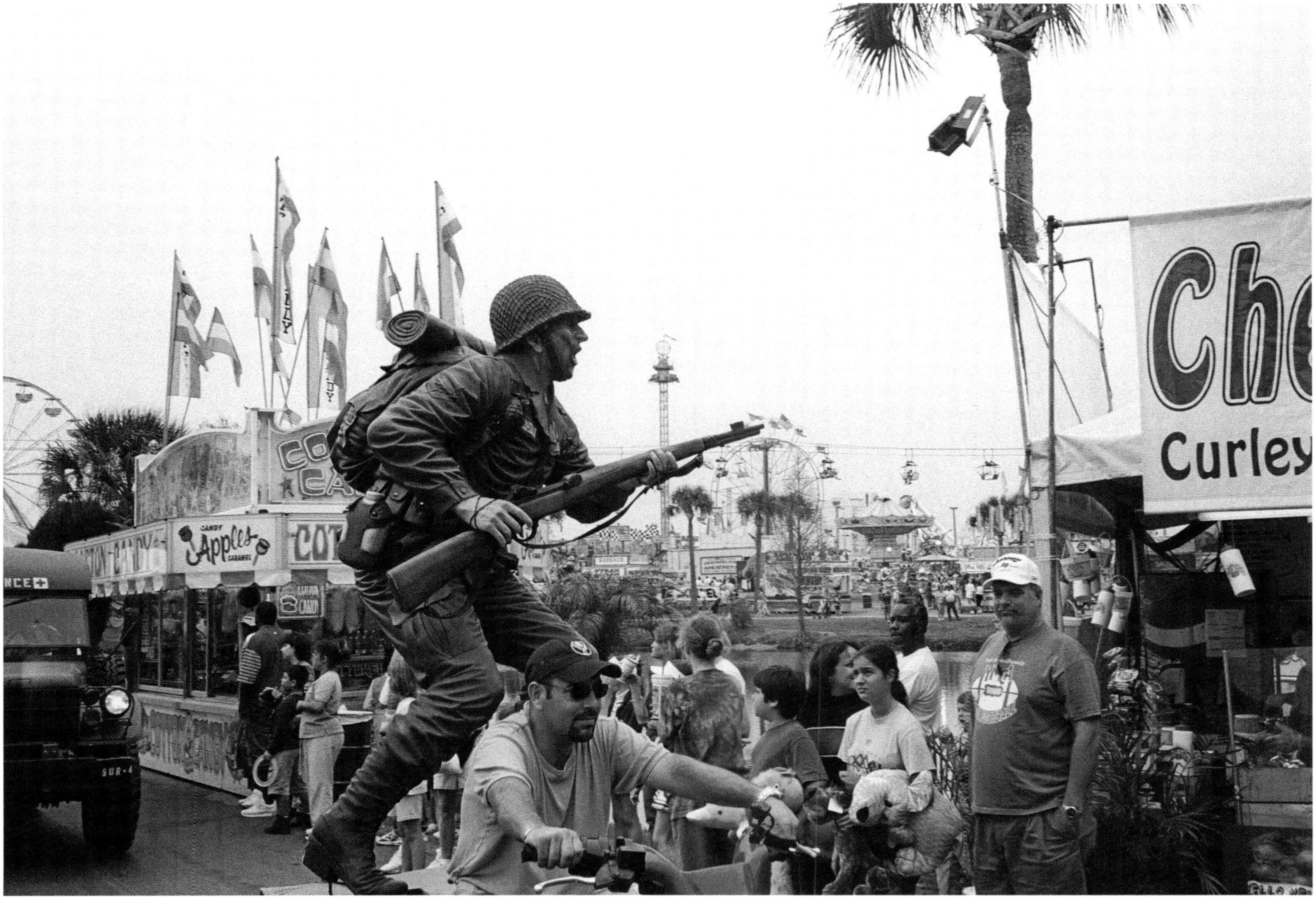
Che
Curley
Apples

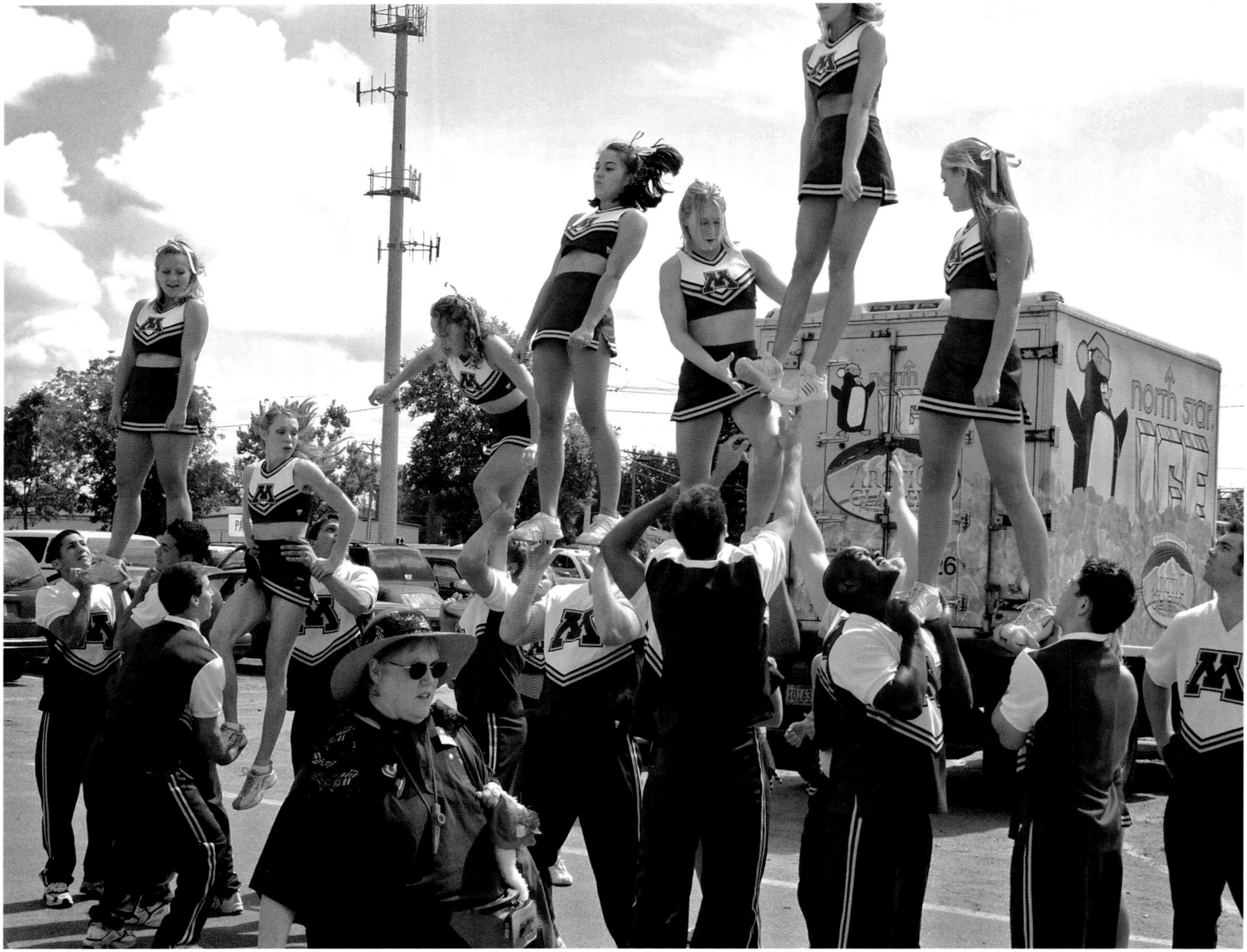
north star.
ICE

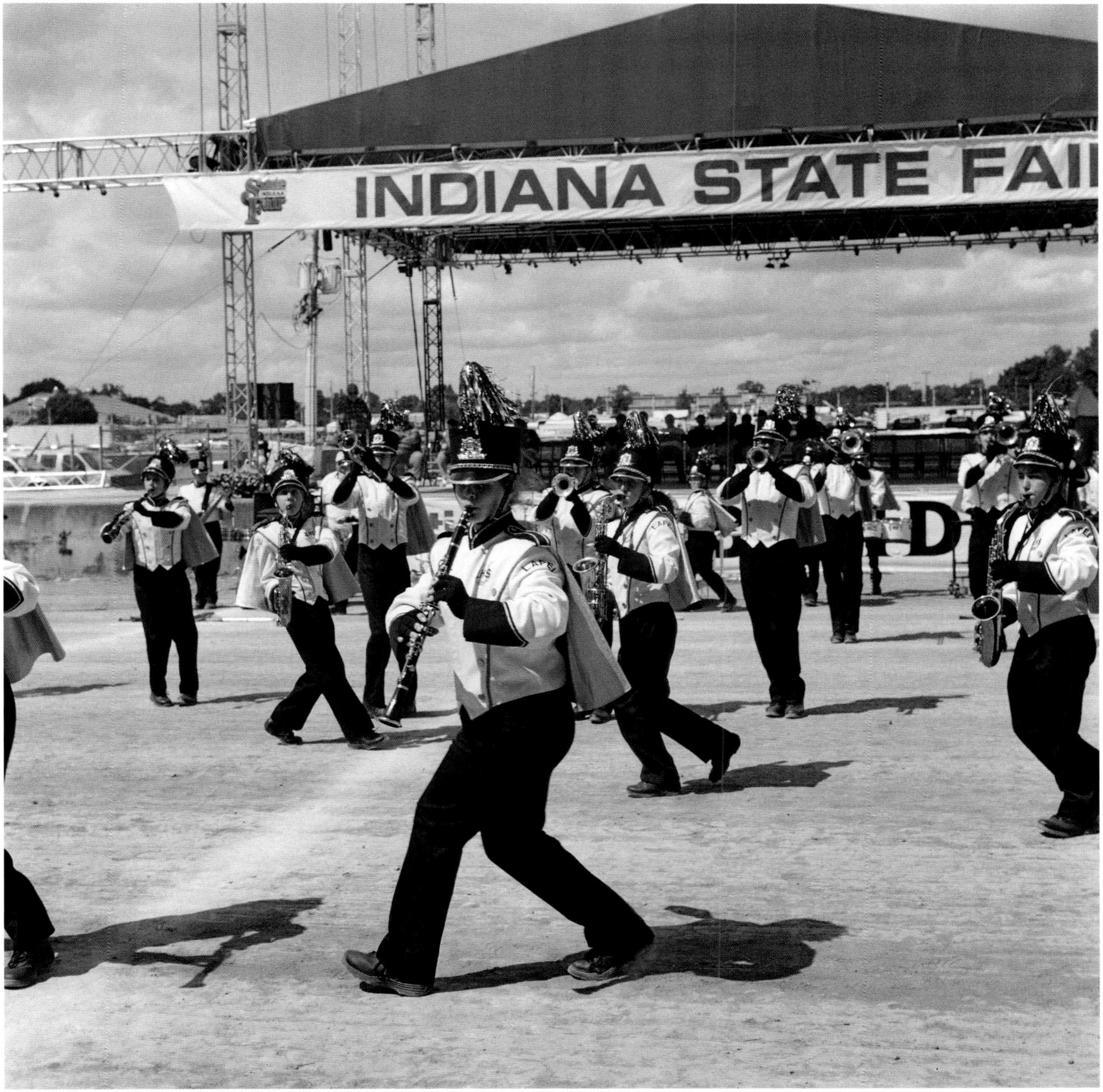
INDIANA STATE FAI

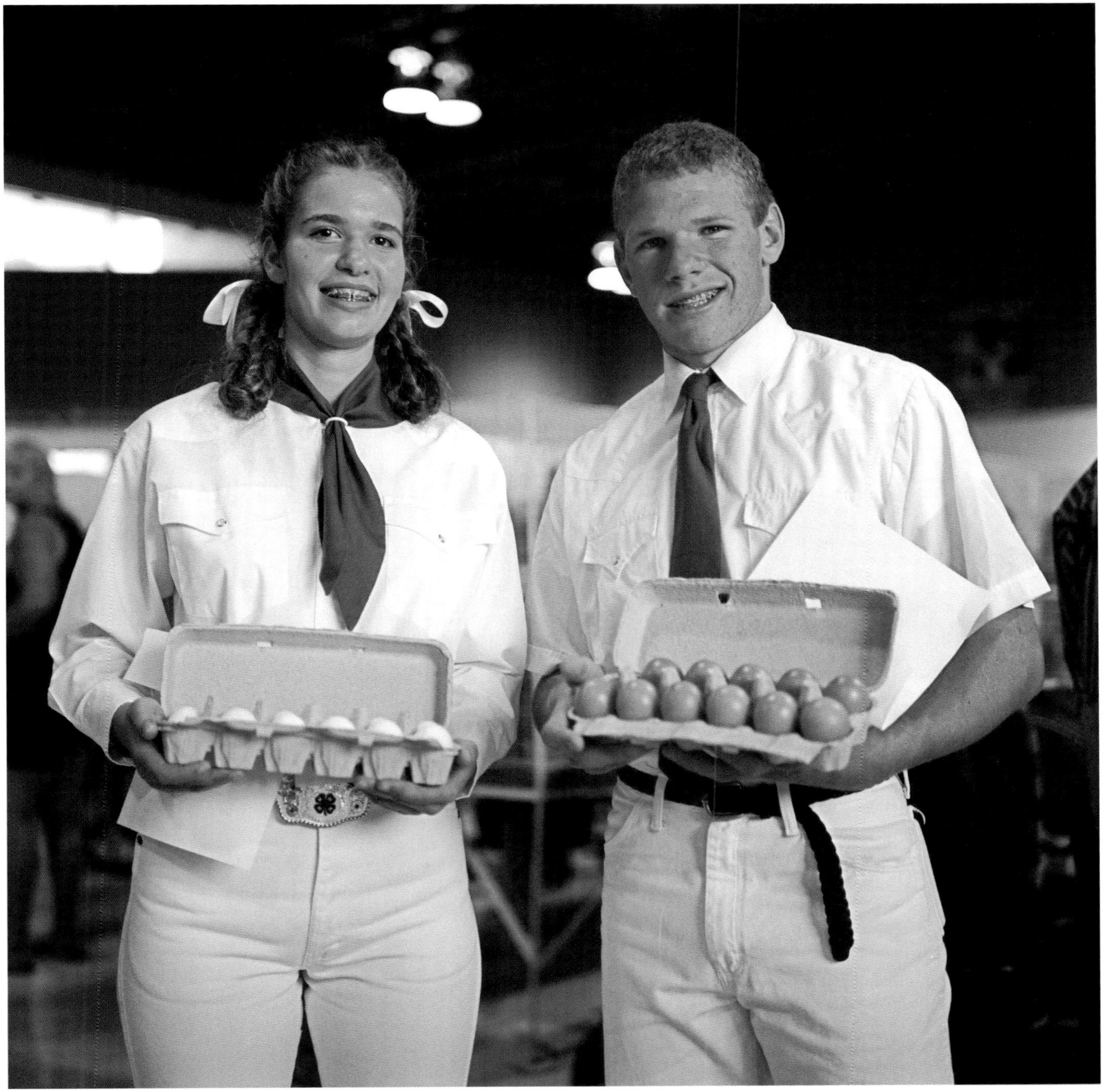

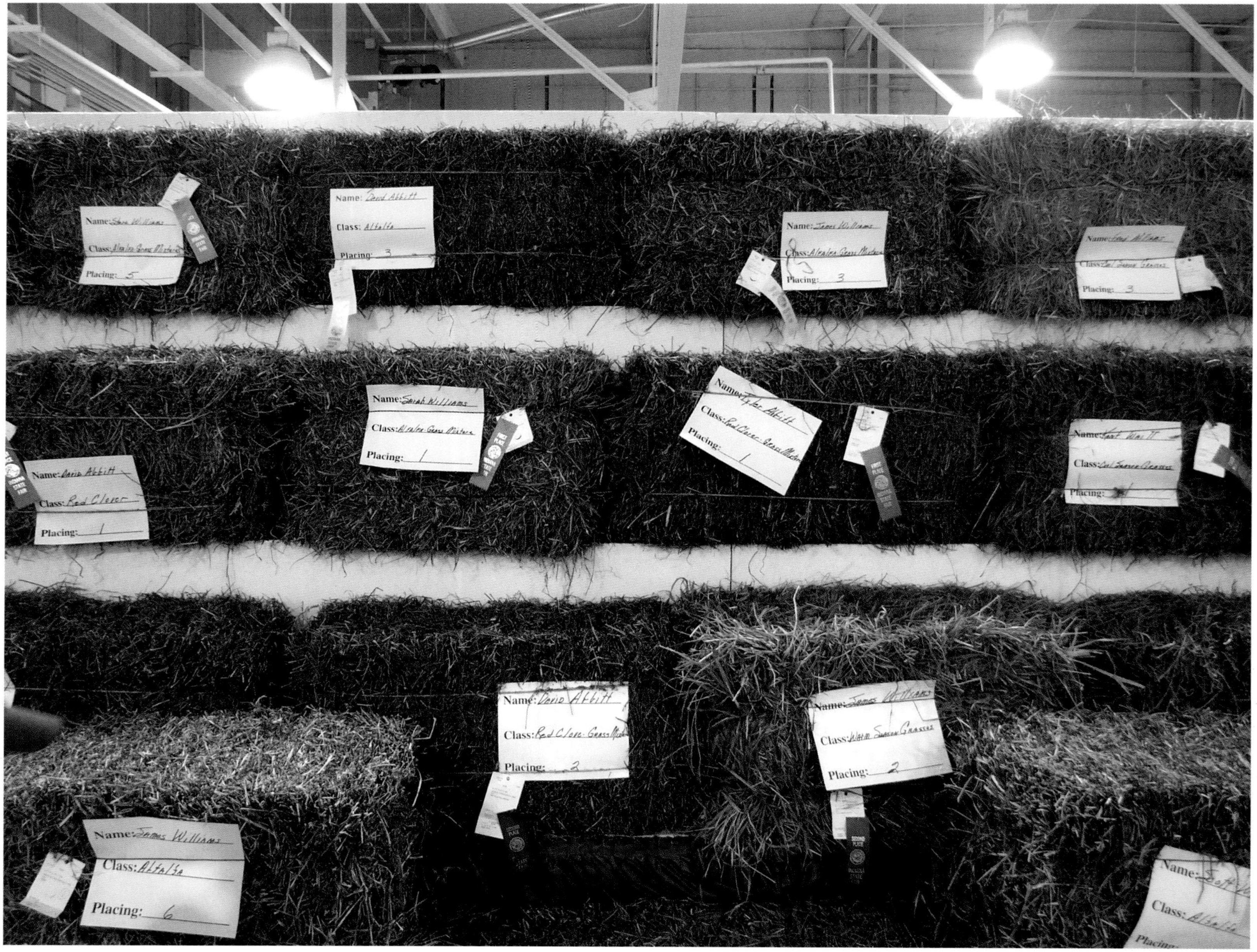
Name: David Abbitt
Class: Alfalfa
Placing: 3
Name: Sarah Williams
Class: Alfalfa-Grass Mixture
Placing: 1
Name: David Abbitt
Class: Red Clover
Placing: 1
Name: James Williams
Class: Alfalfa
Placing: 6

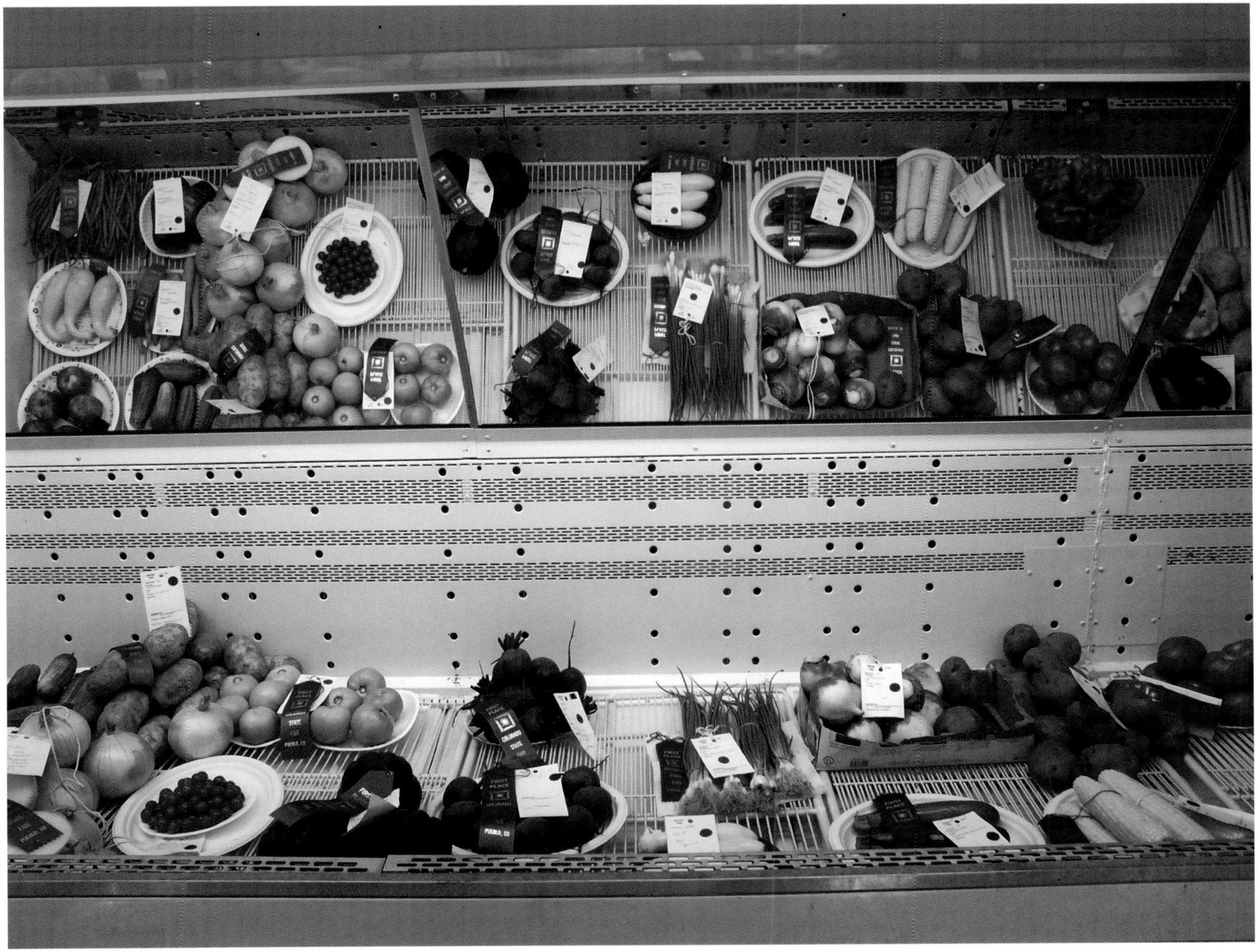

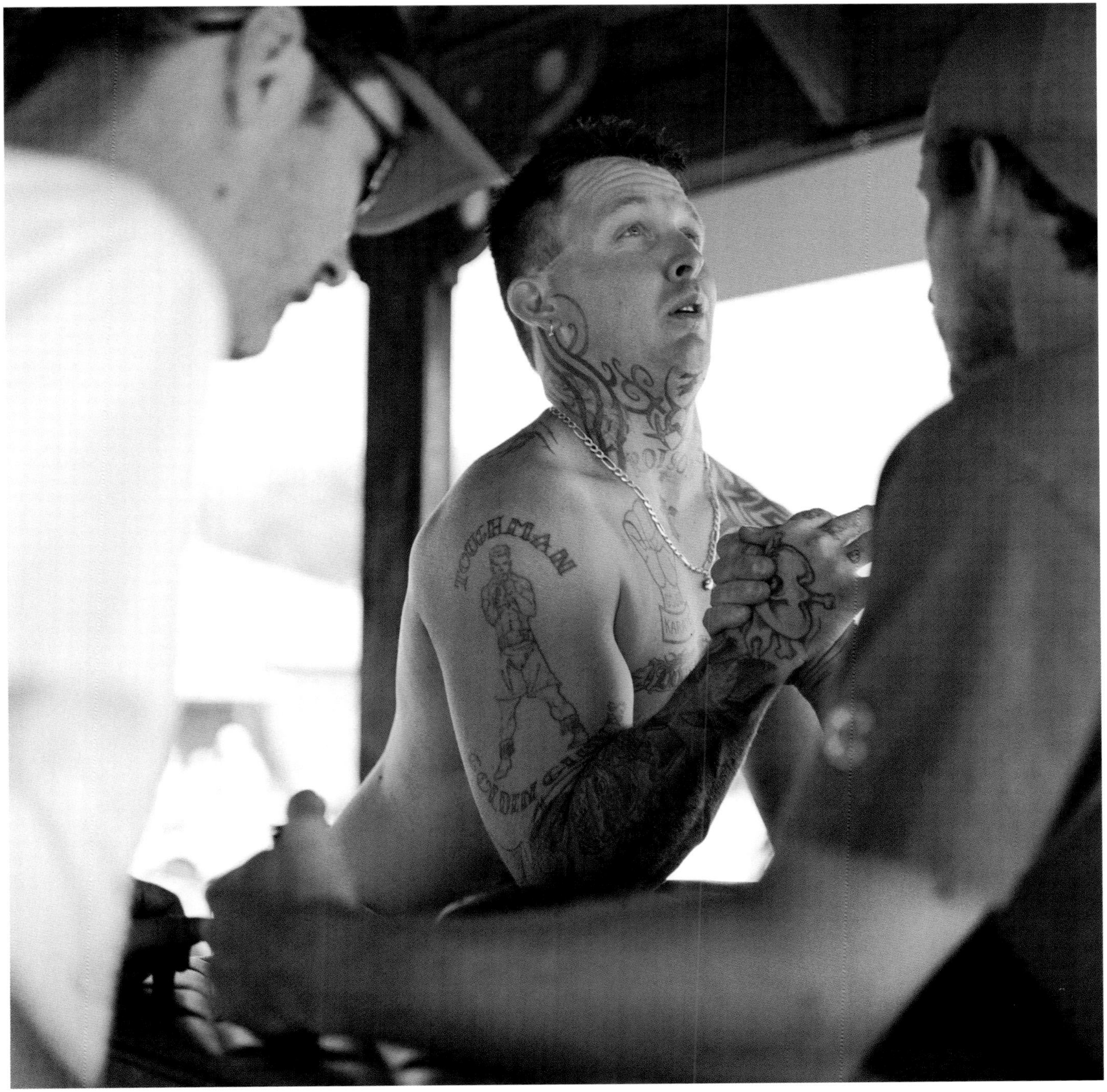
TOUGHMAN

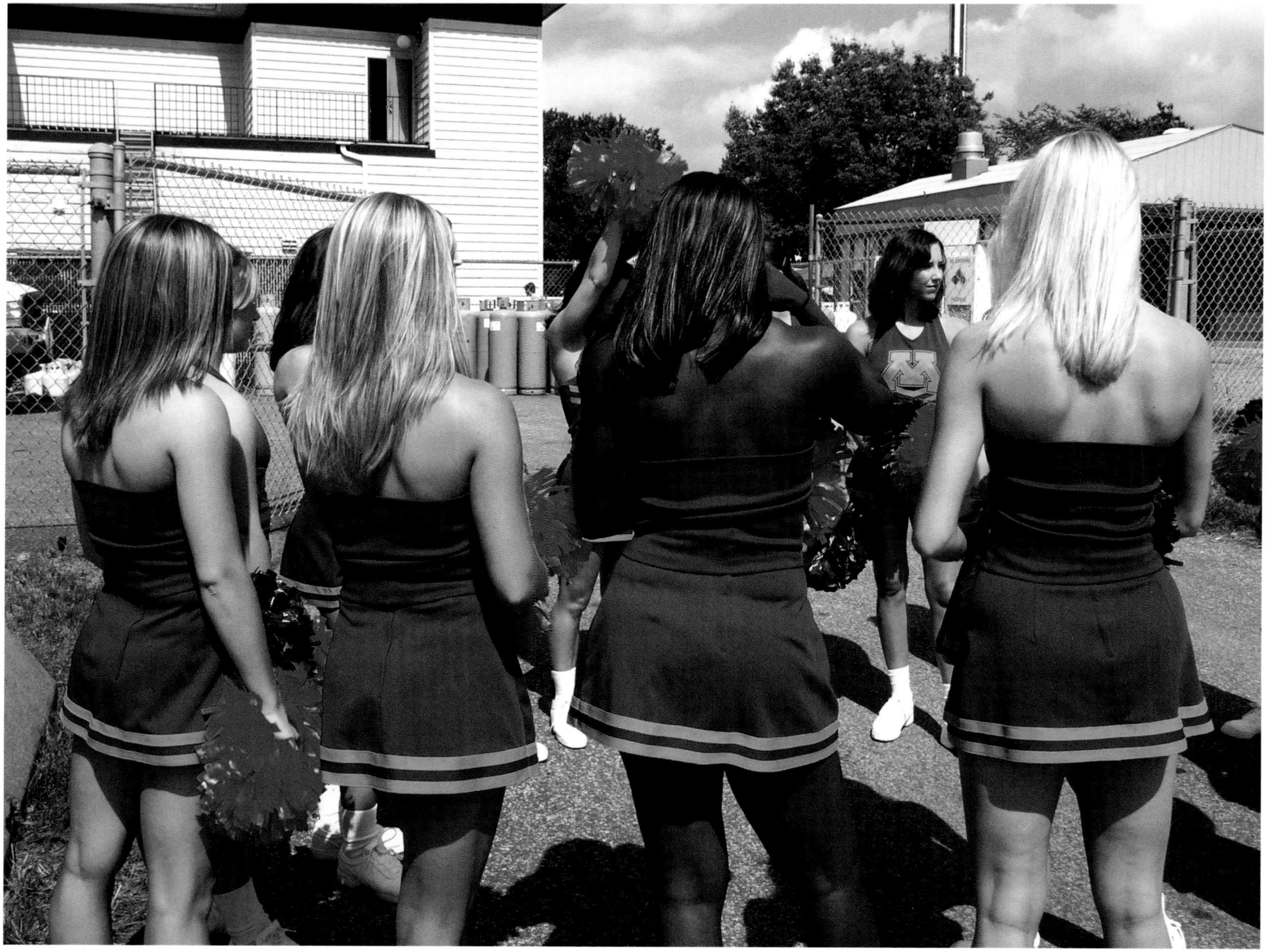

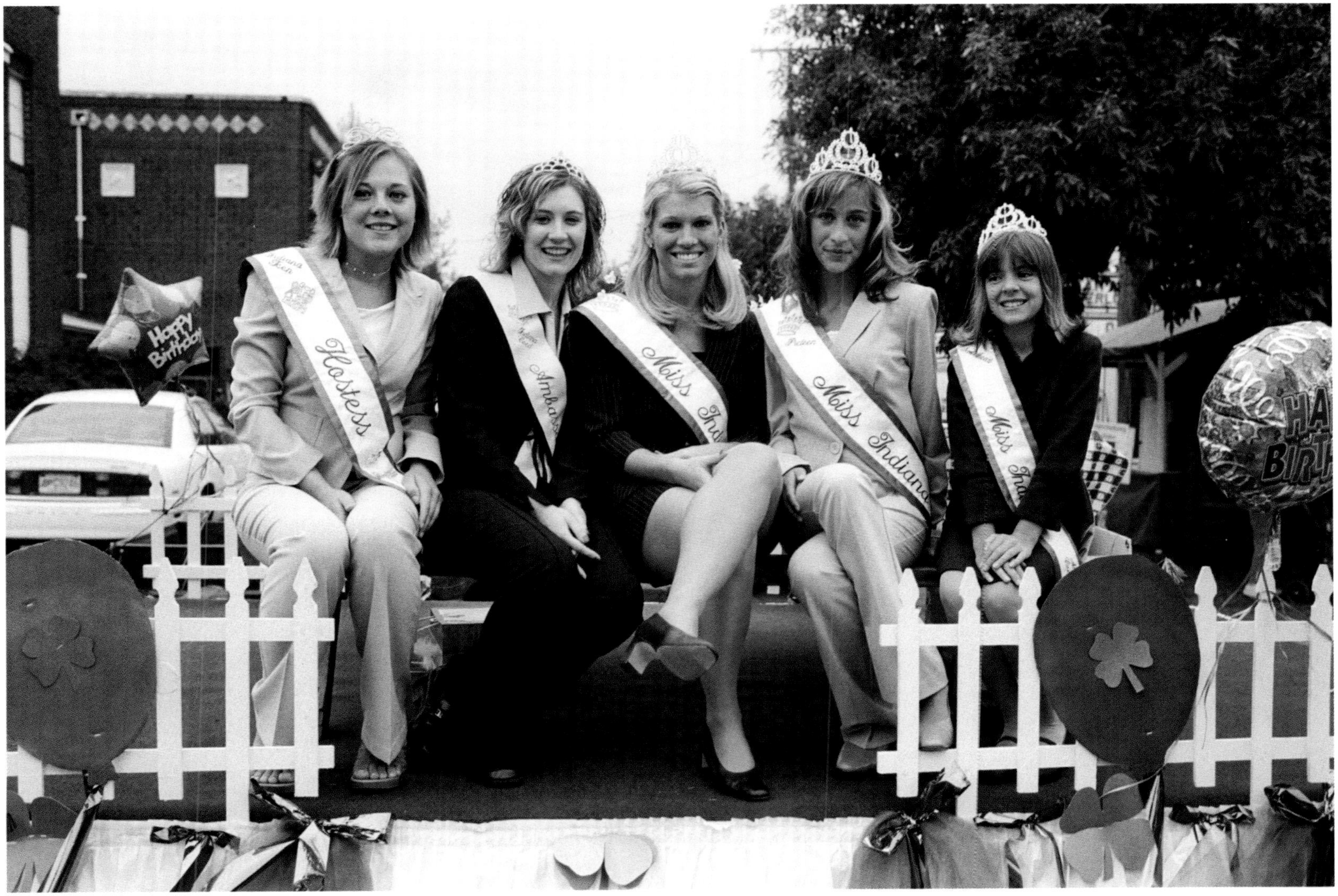
Happy Birthday
Hostess
Miss Indiana

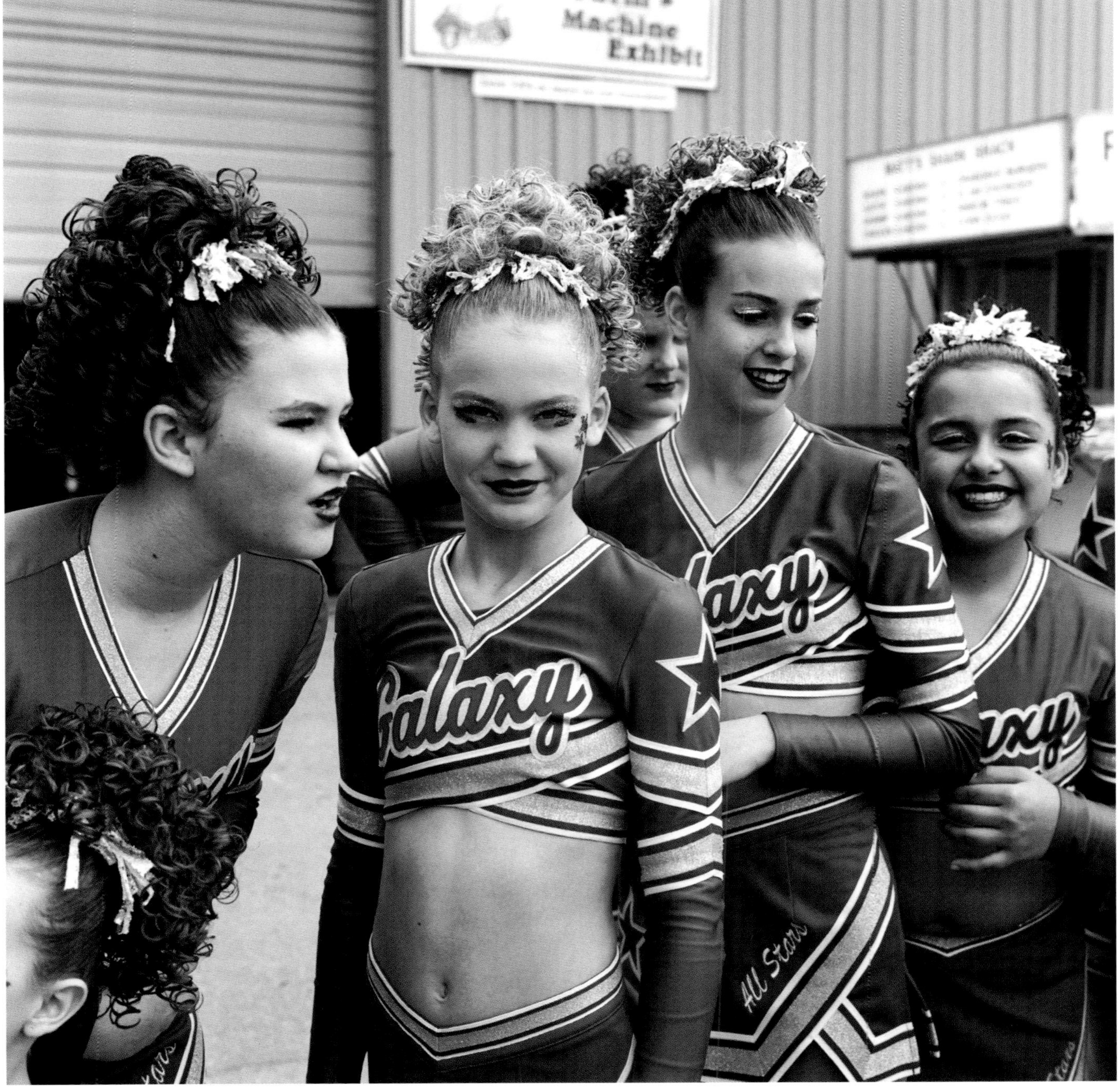
Farm &
Machine
Exhibit
Galaxy
All Stars

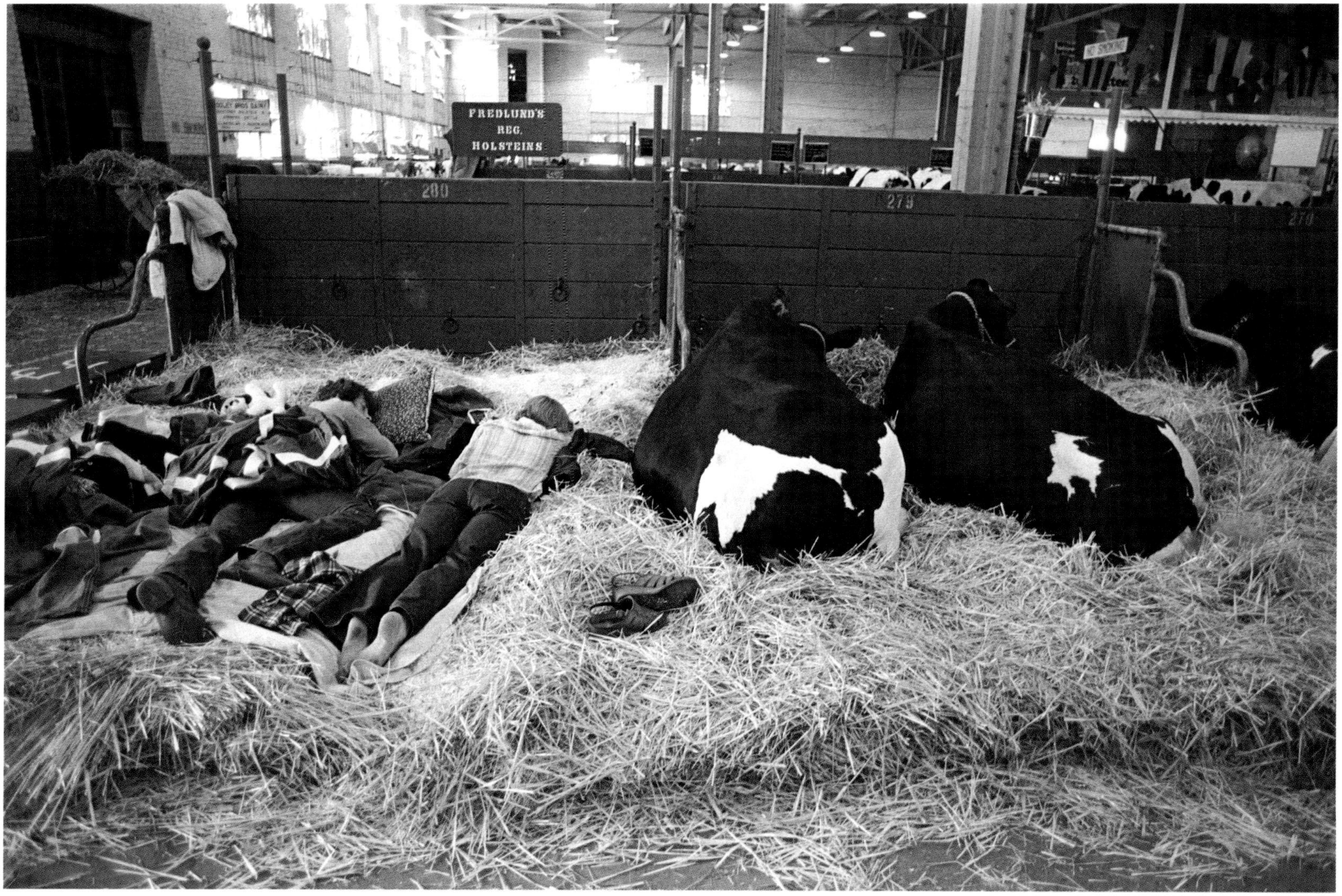
FREDLUND'S
REG.
HOLSTEINS
280
279

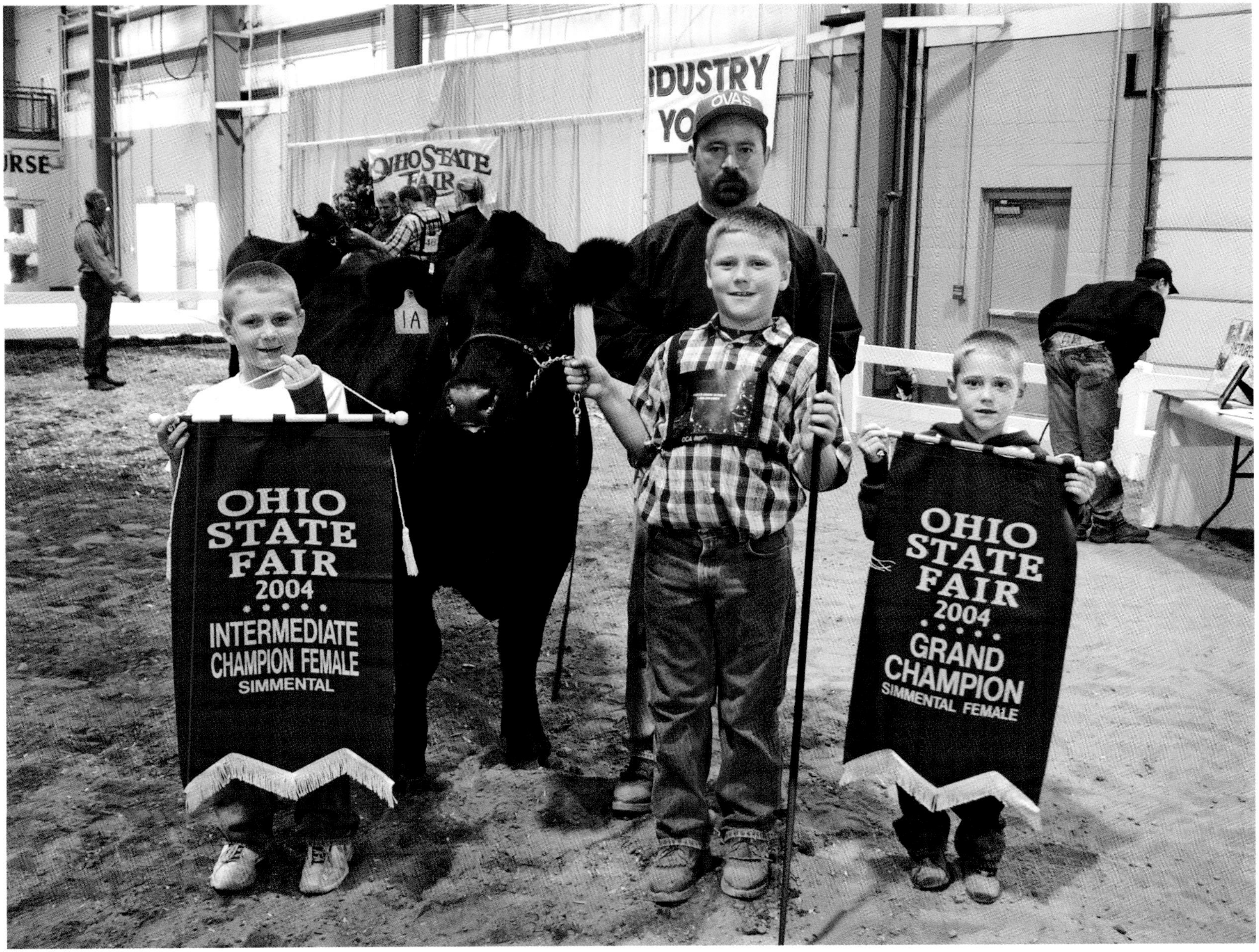
DUSTRY
YO
OVAS
OHIO STATE FAIR
1A
OHIO
STATE
FAIR
2004
INTERMEDIATE
CHAMPION FEMALE
SIMMENTAL
OHIO
STATE
FAIR
2004
GRAND
CHAMPION
SIMMENTAL FEMALE

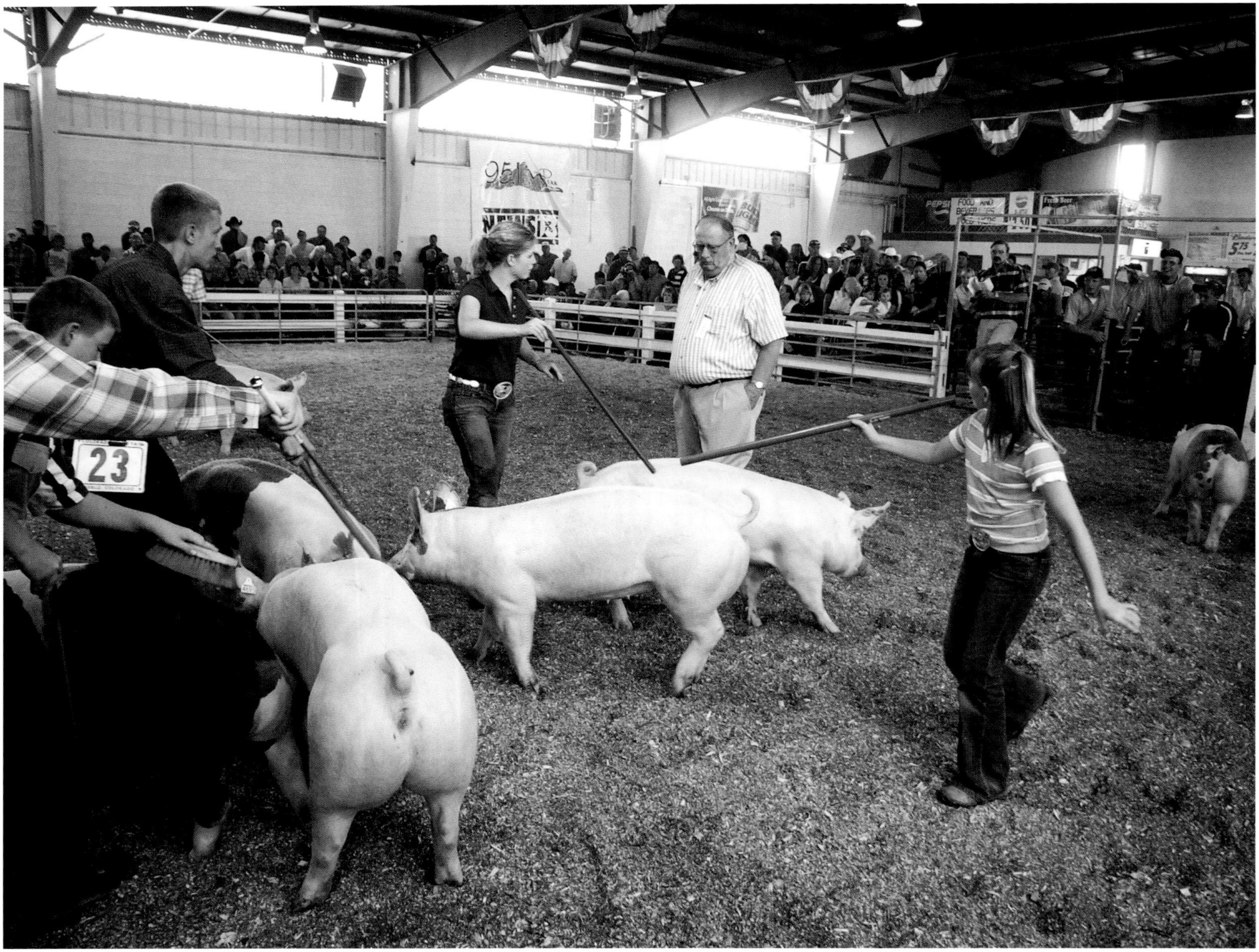
23

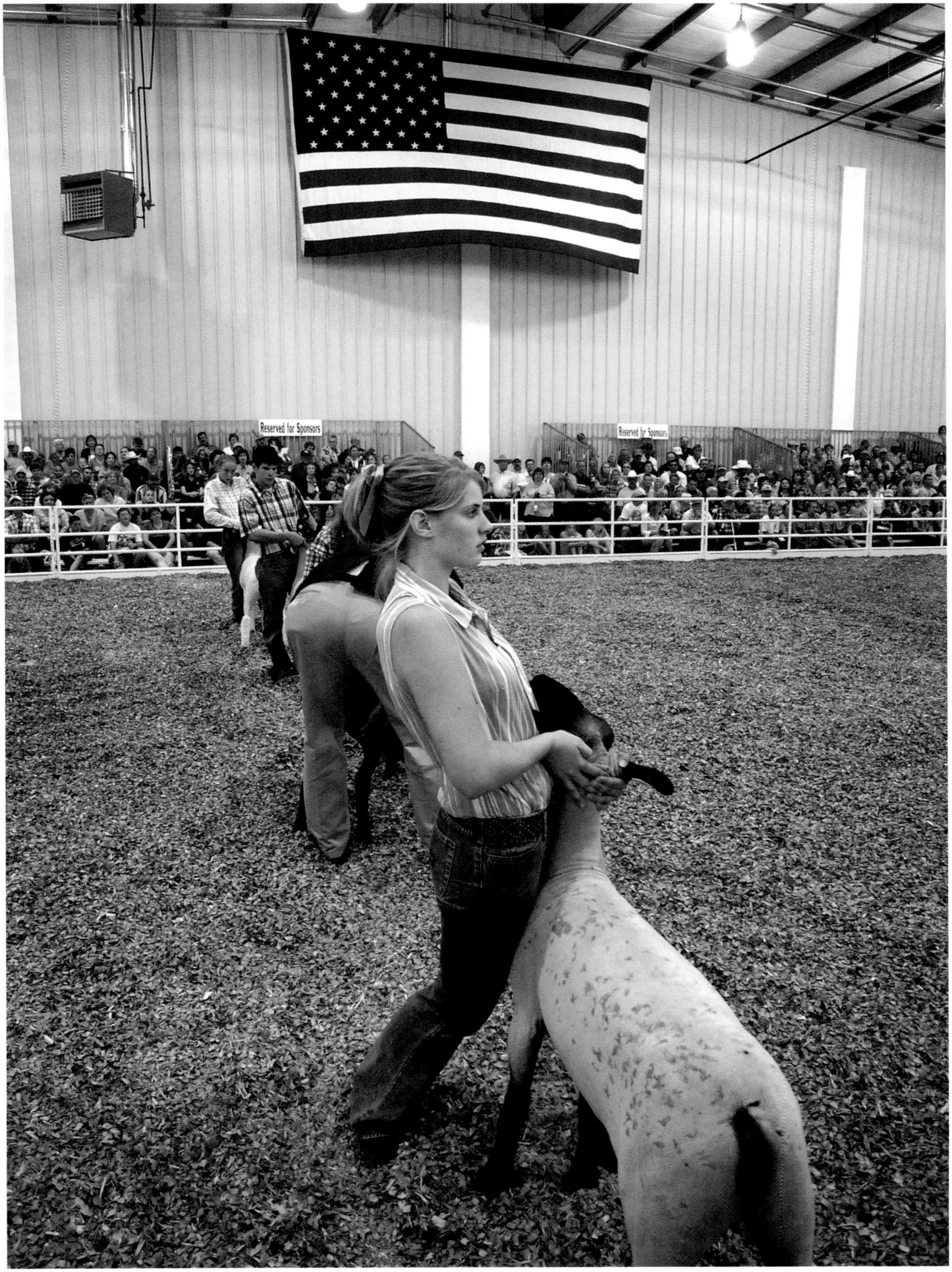
Reserved for Sponsors
Reserved for Sponsors

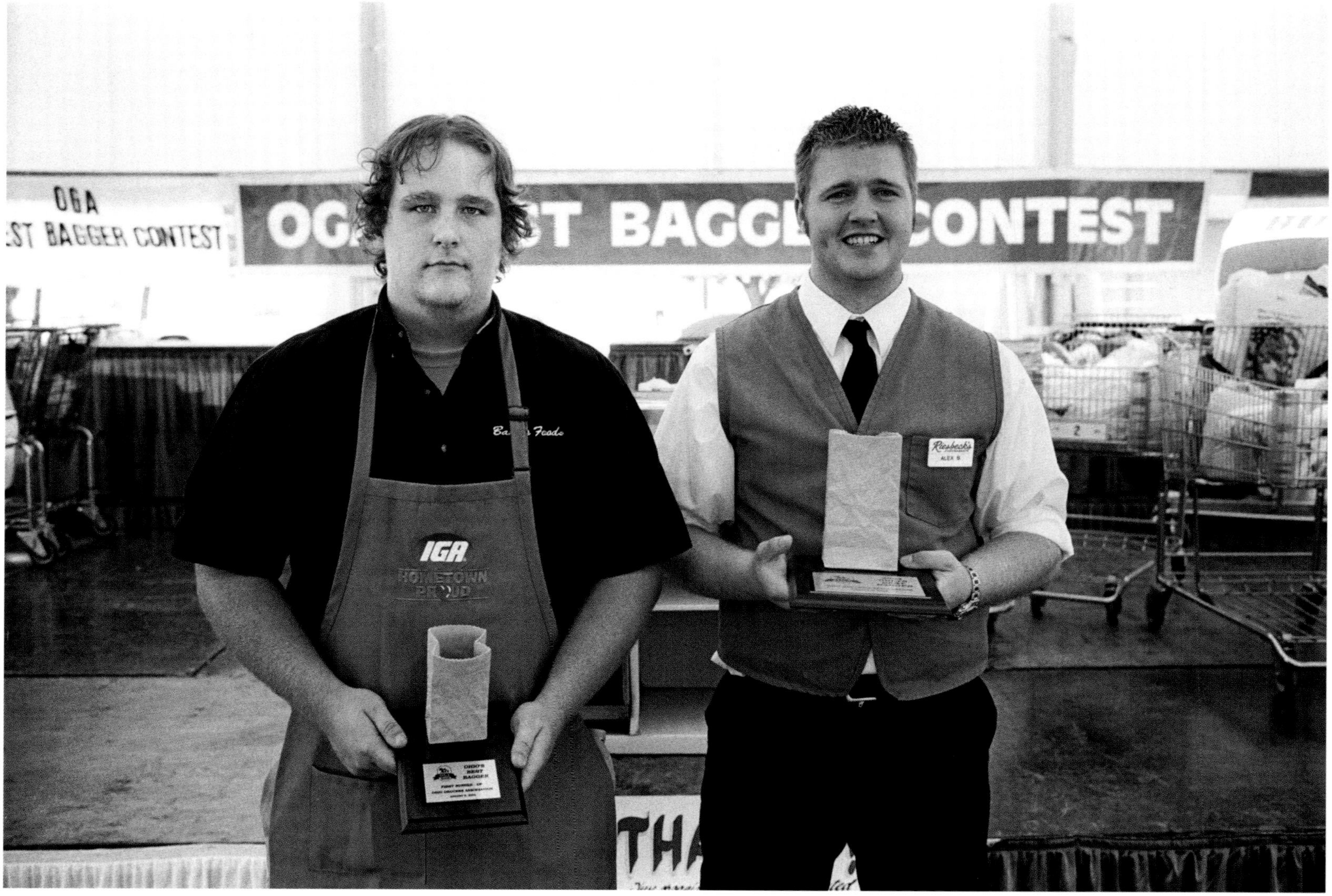
OGA
EST BAGGER CONTEST
CONTEST
IGA
HOMETOWN PROUD
Riesbeck's

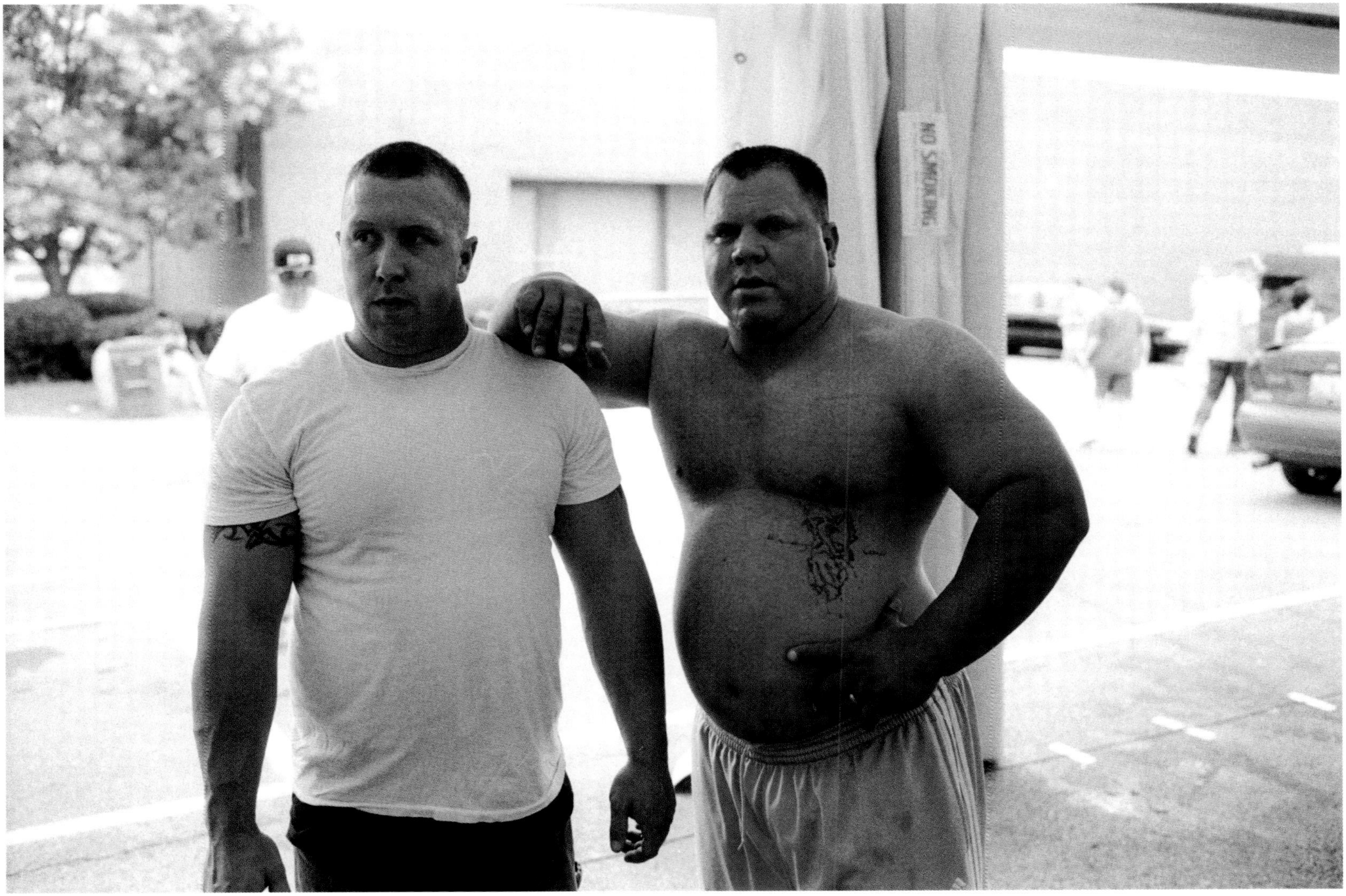
NO SMOKING

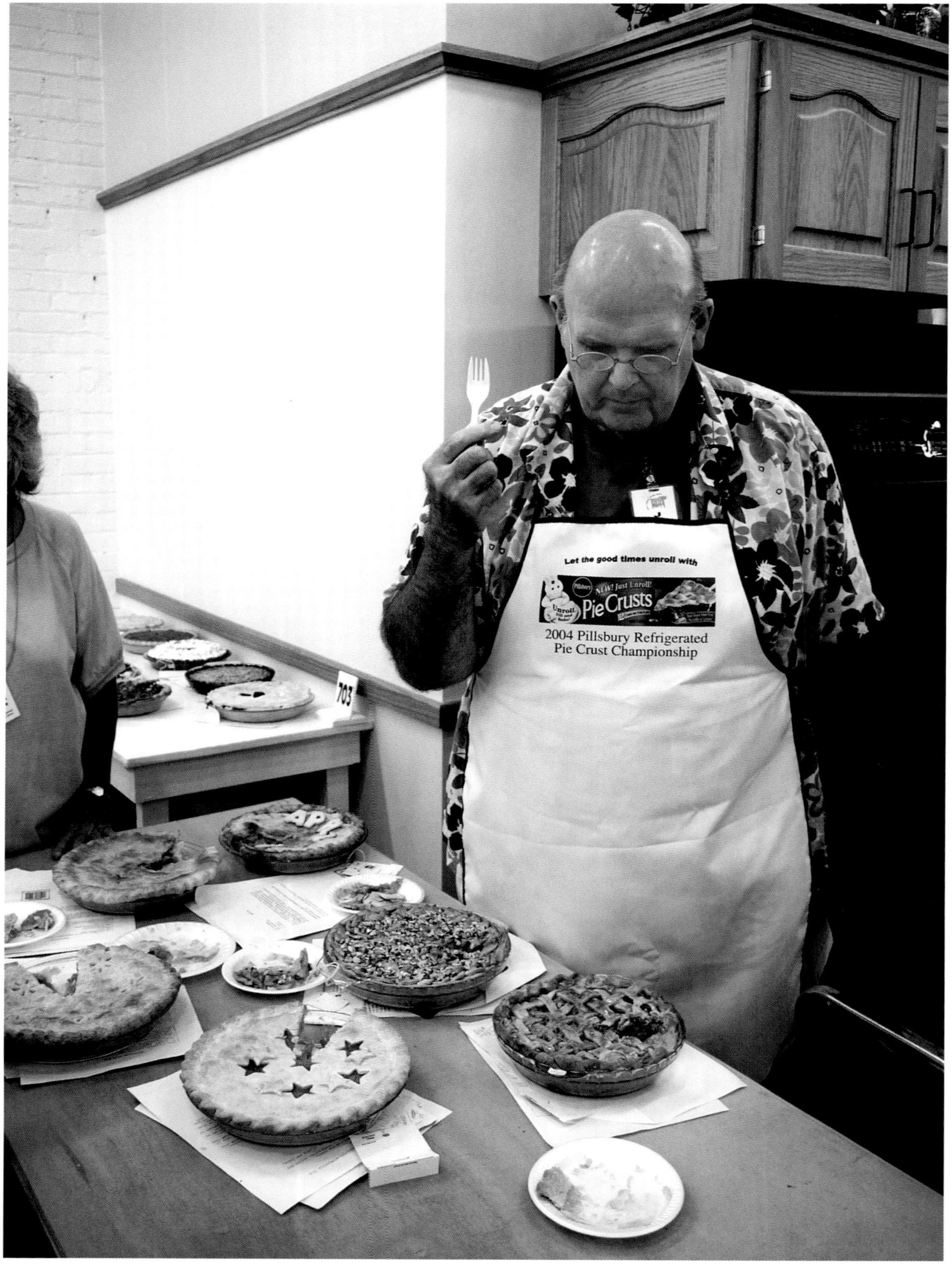
Let the good times unroll with
Pie Crusts
2004 Pillsbury Refrigerated
Pie Crust Championship
703

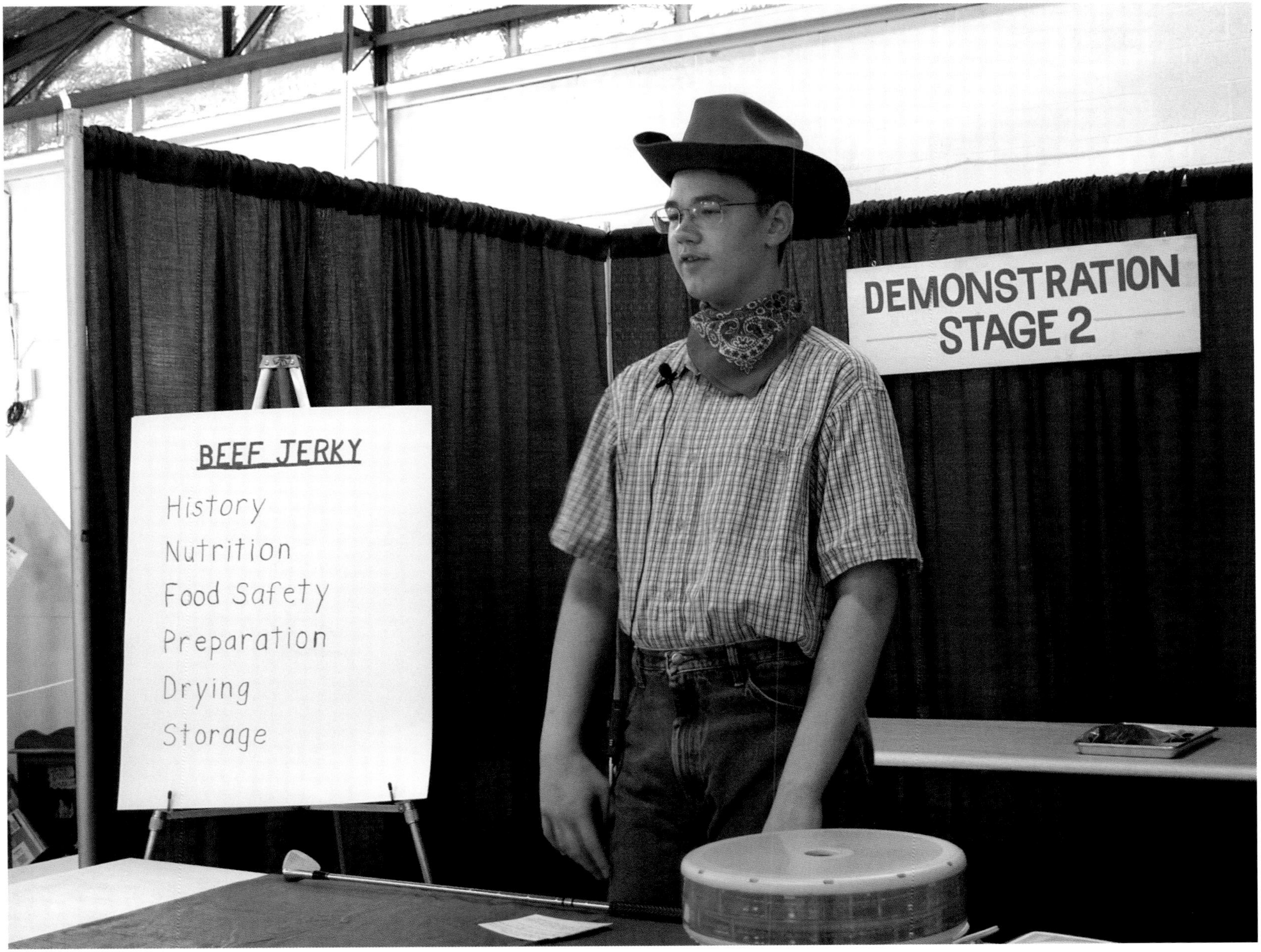
BEEF JERKY
History
Nutrition
Food Safety
Preparation
Drying
Storage
DEMONSTRATION
STAGE 2

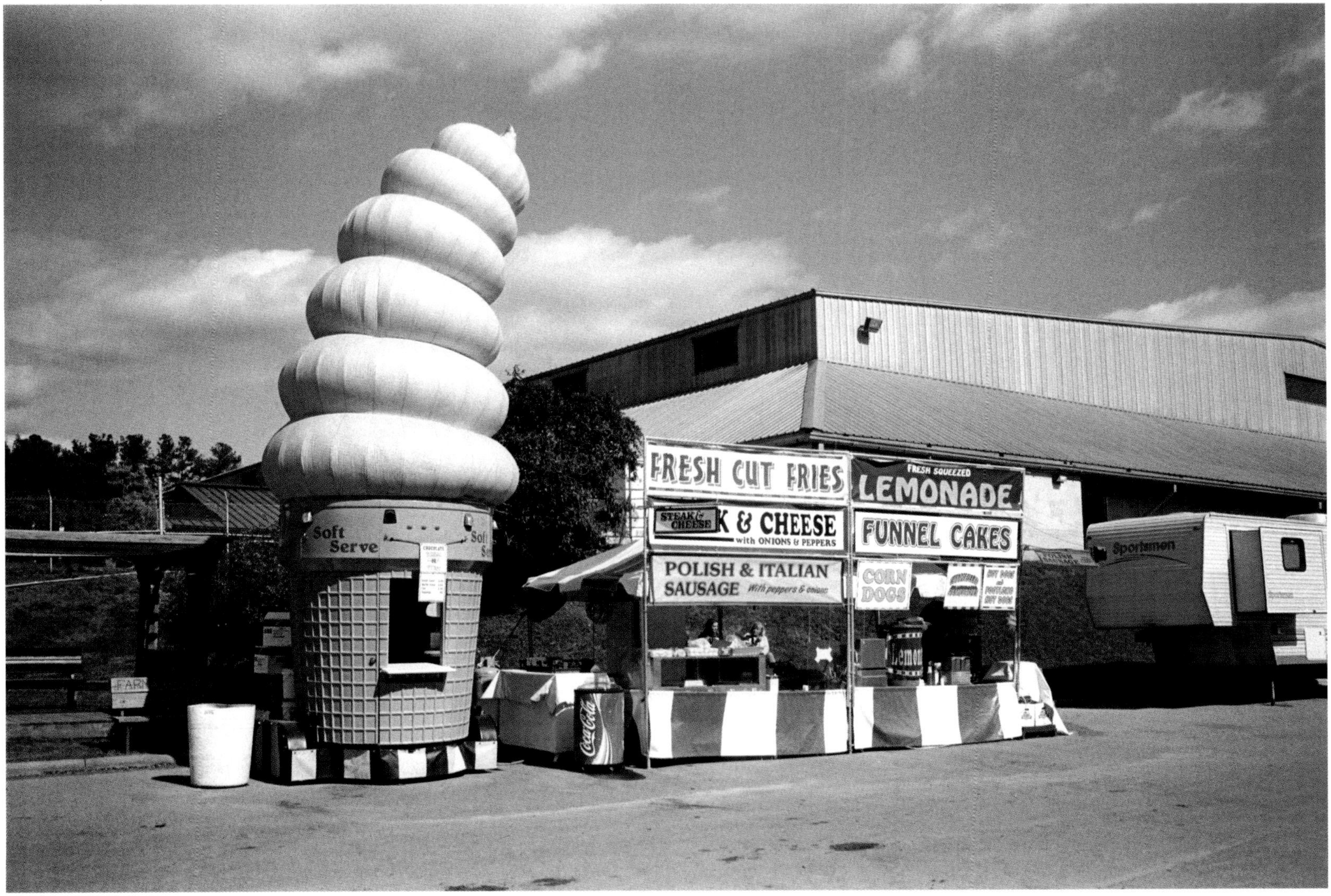
Soft Serve
FRESH CUT FRIES
FRESH SQUEEZED
LEMONADE
STEAK & CHEESE
K & CHEESE
with ONIONS & PEPPERS
FUNNEL CAKES
POLISH & ITALIAN
SAUSAGE With peppers & onions
CORN DOGS
Coca-Cola
Sportsmen

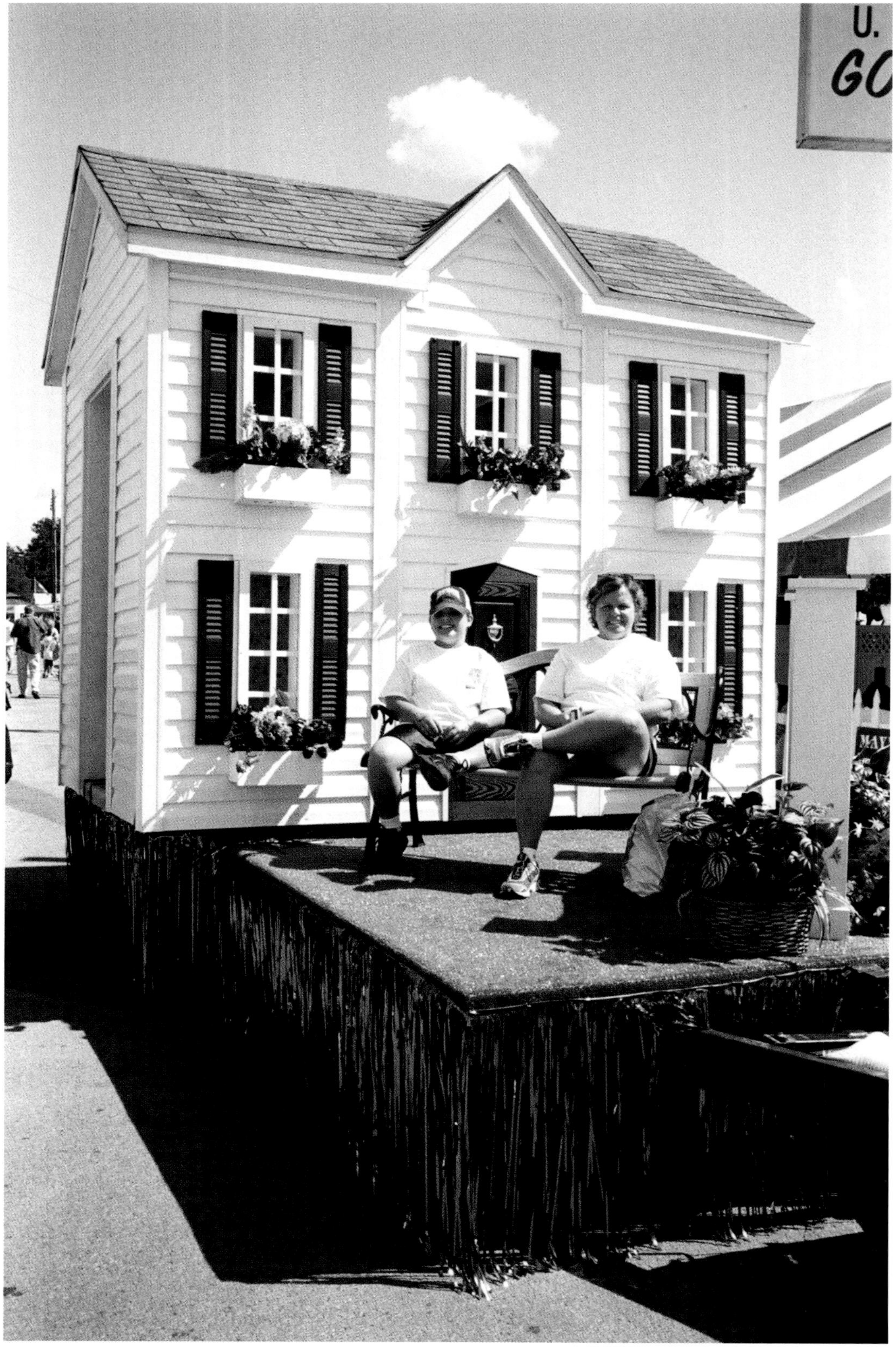

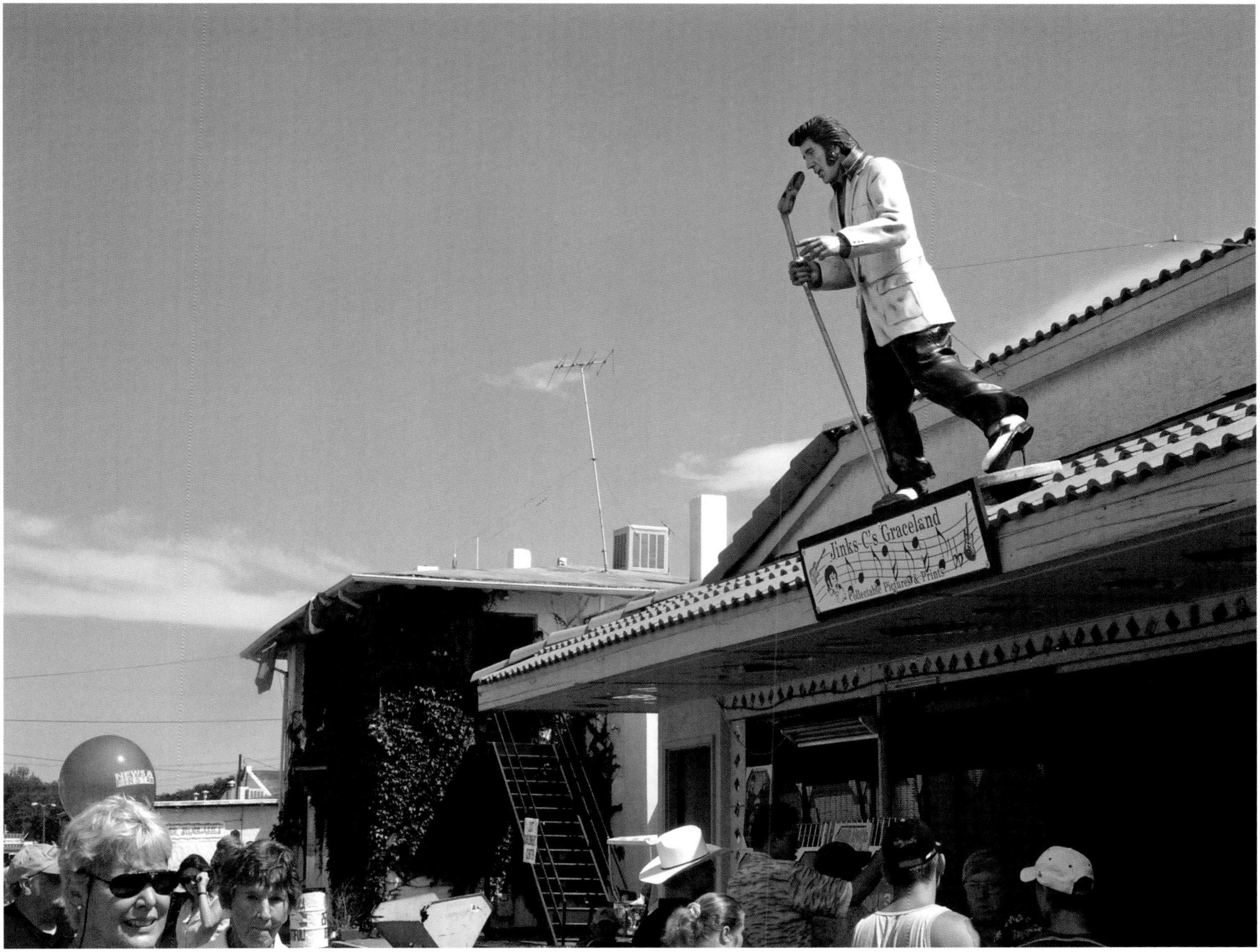
Jinks-C's Graceland
Collectable Pictures & Prints

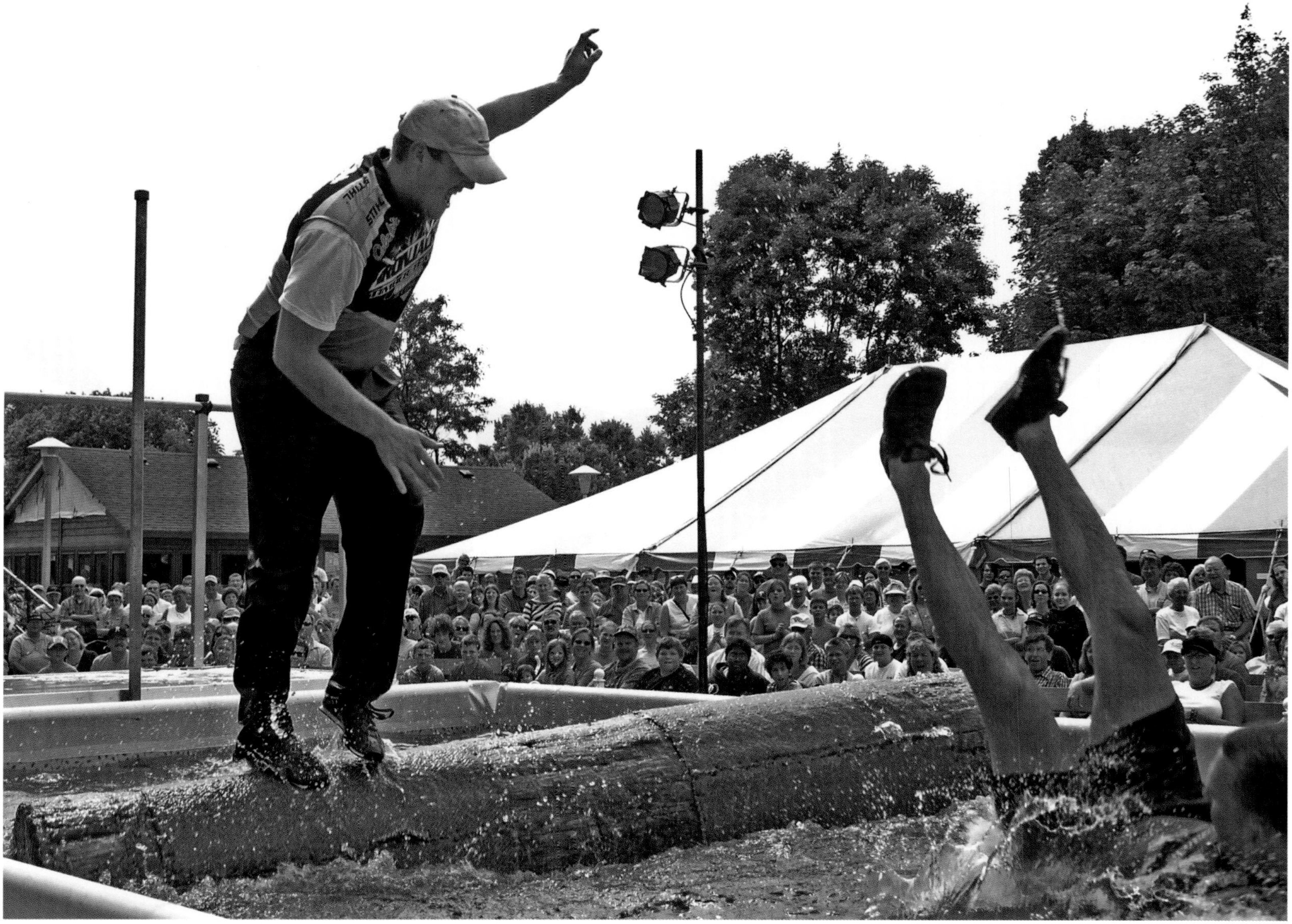

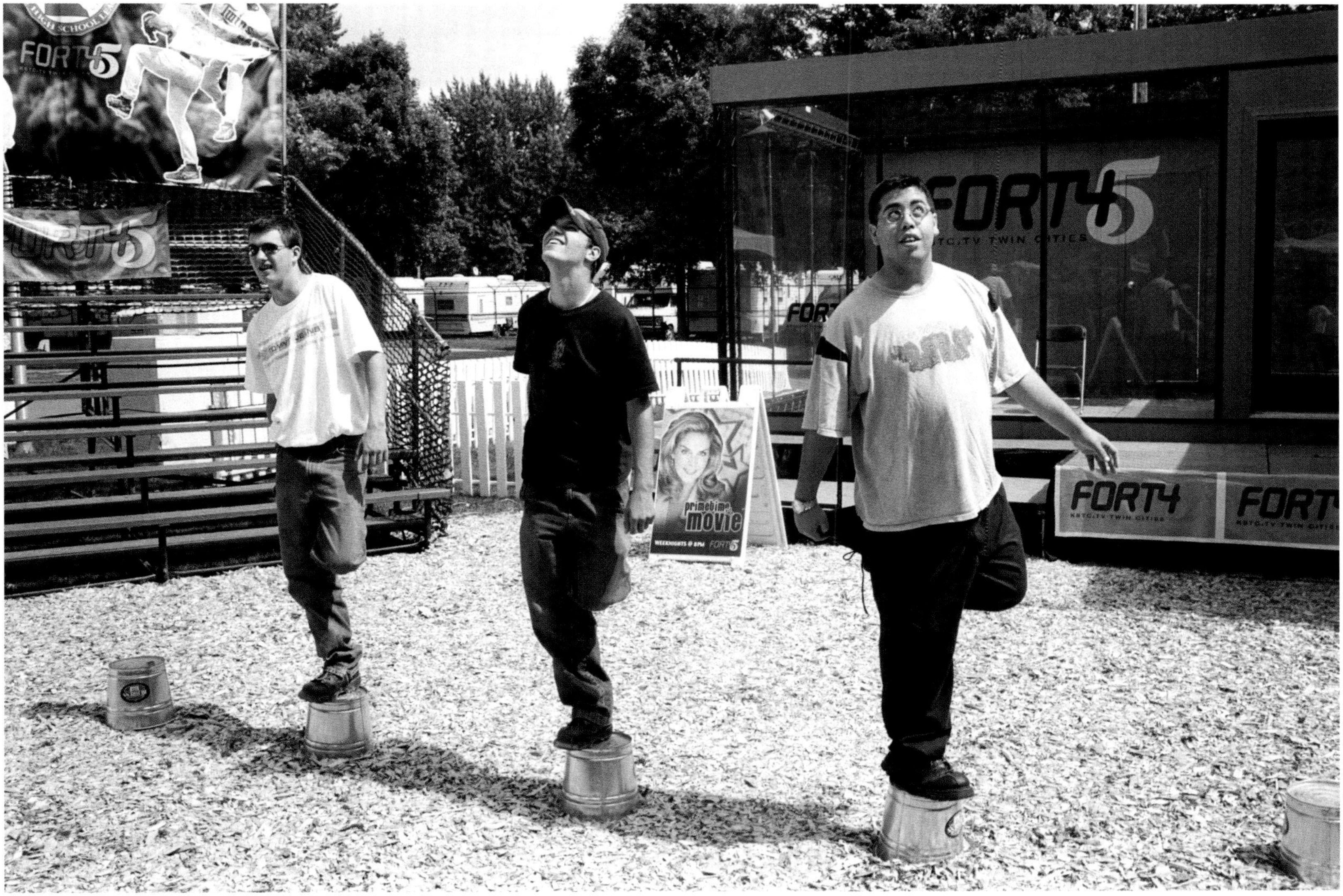
primetime movie
WEEKNIGHTS @ 8PM
KSTC-TV TWIN CITIES

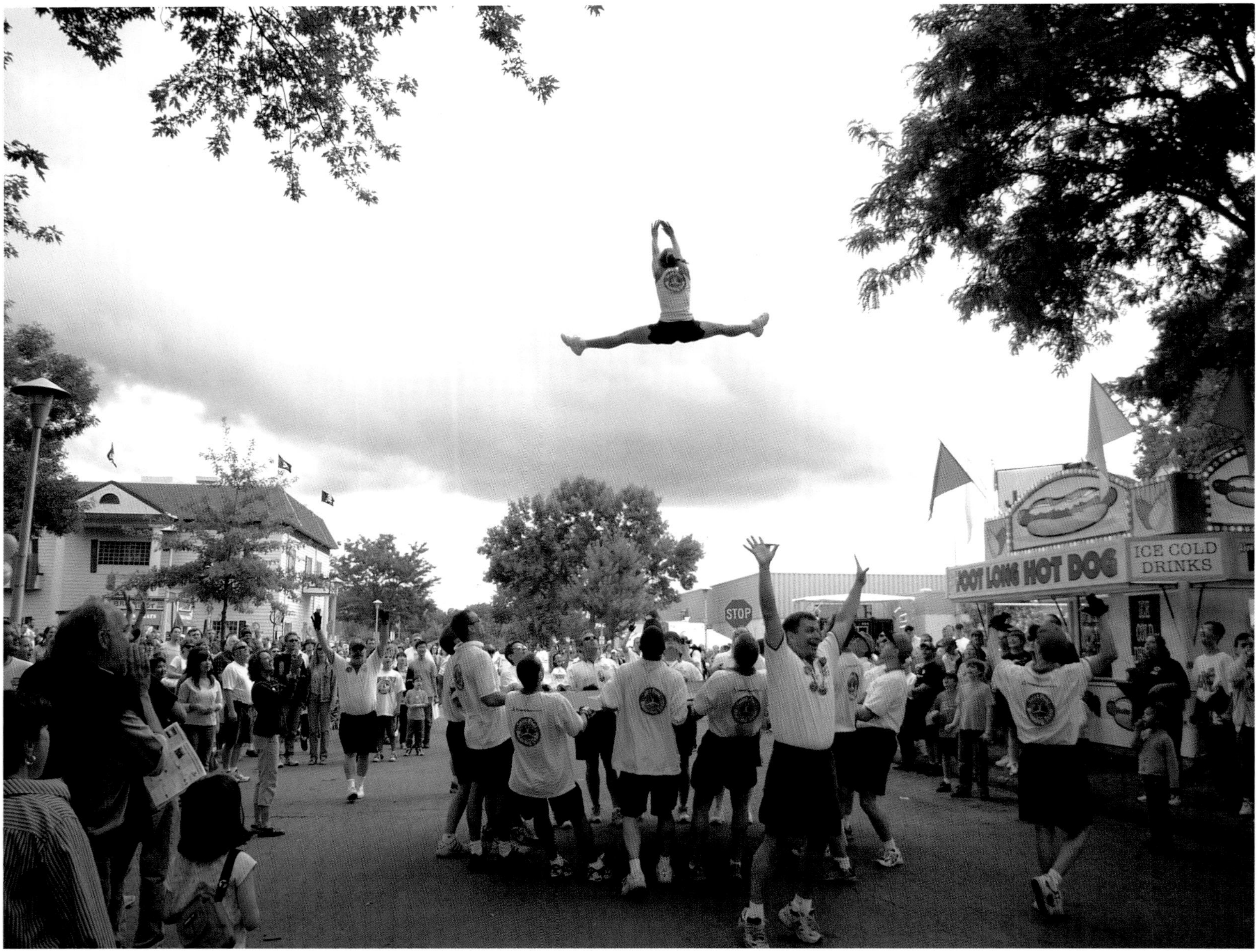
ICE COLD
DRINKS
STOP

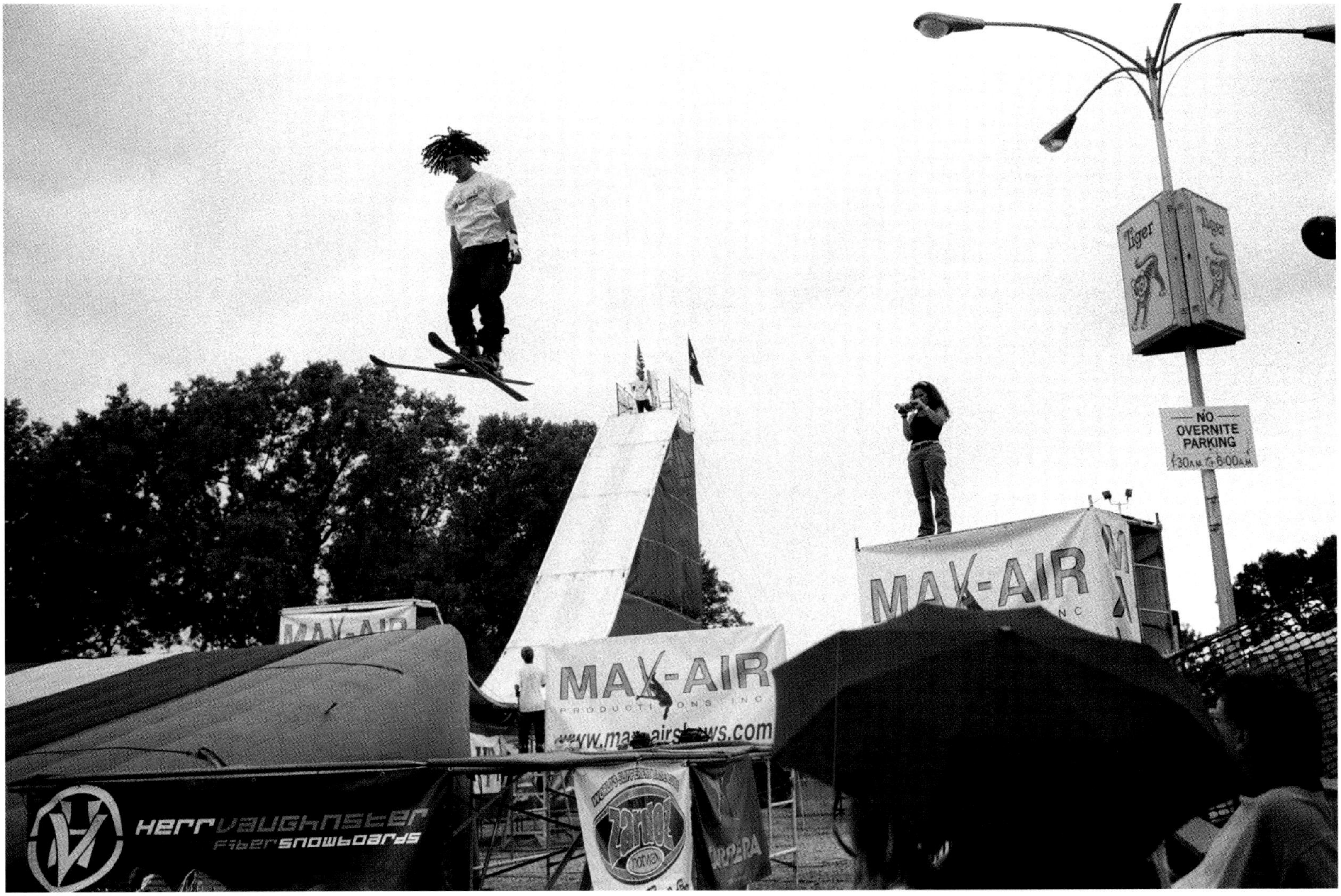
Tiger
Tiger
NO
OVERNITE
PARKING
MAX-AIR
MAX-AIR
herrvaughnster
fibersnowboards

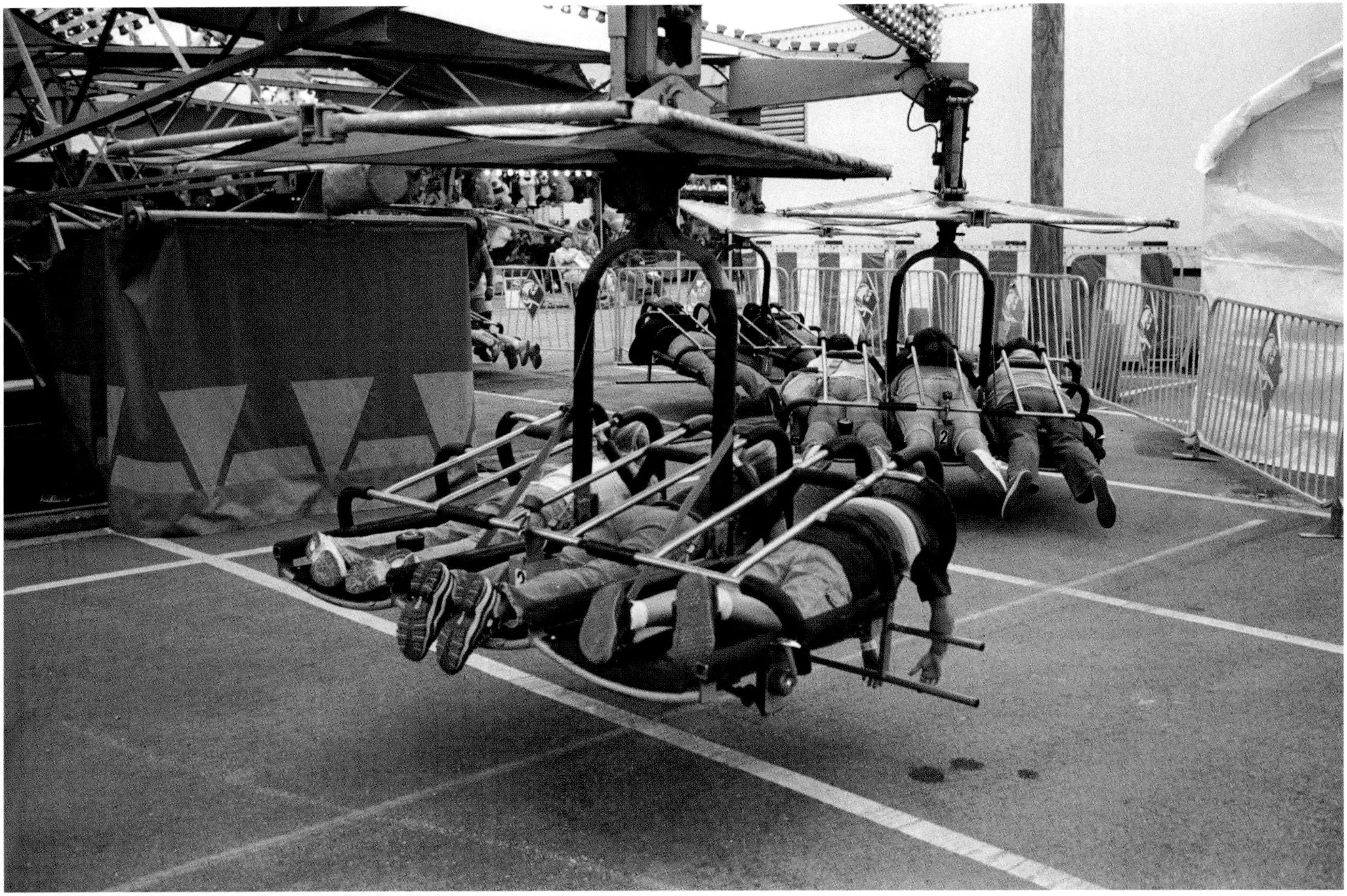

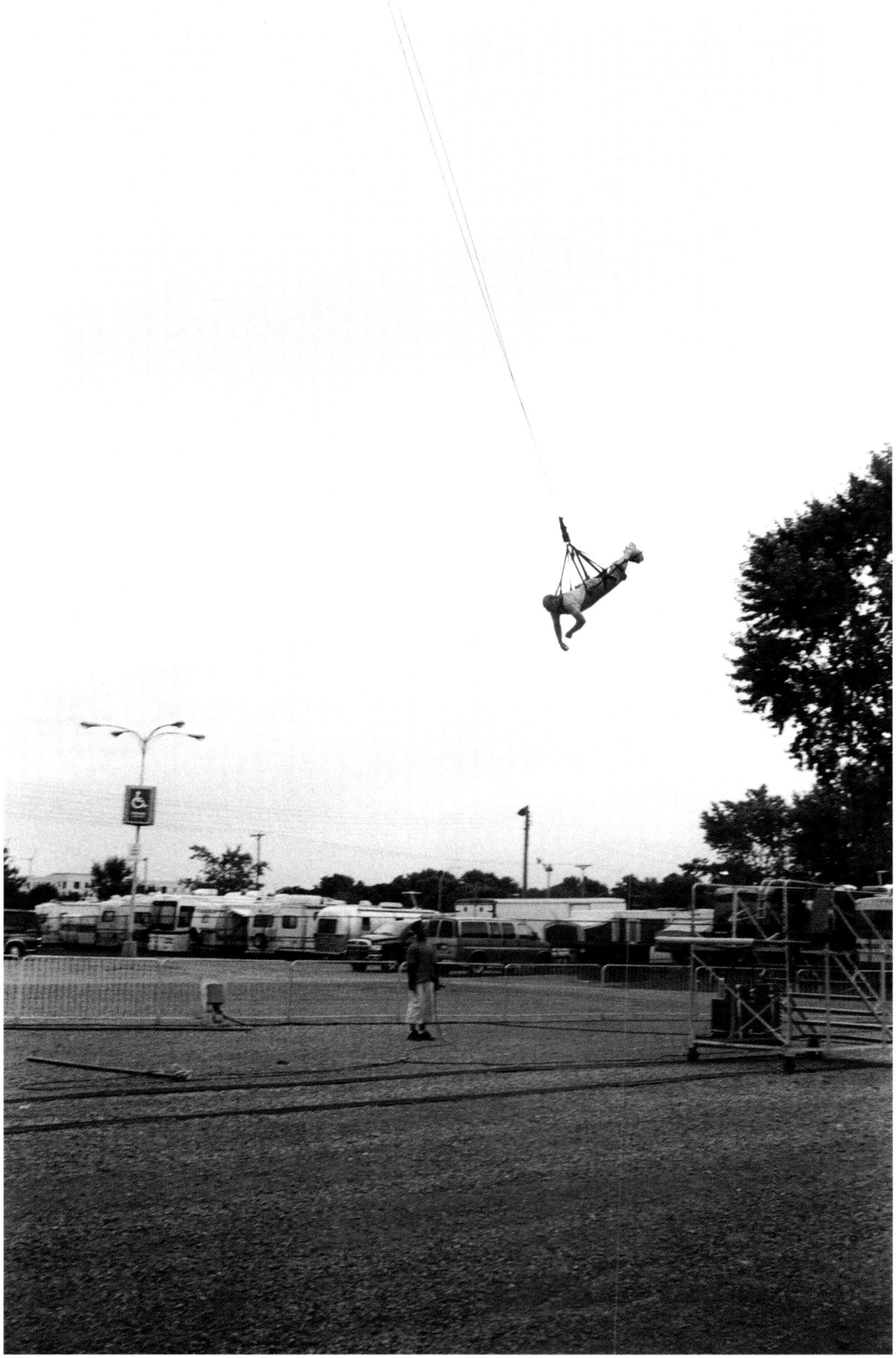

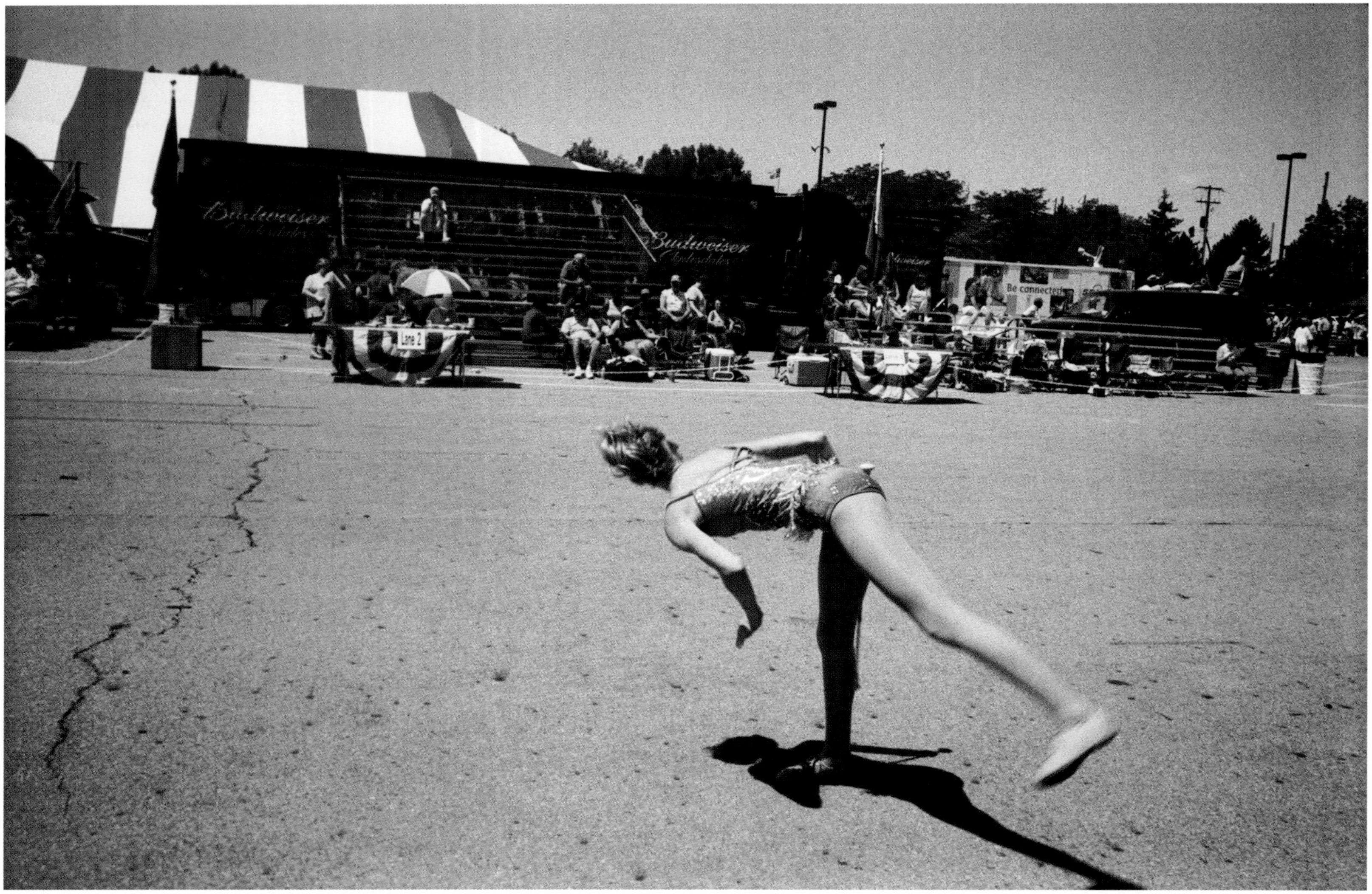
Budweiser
Budweiser
Lane 2
Be connected

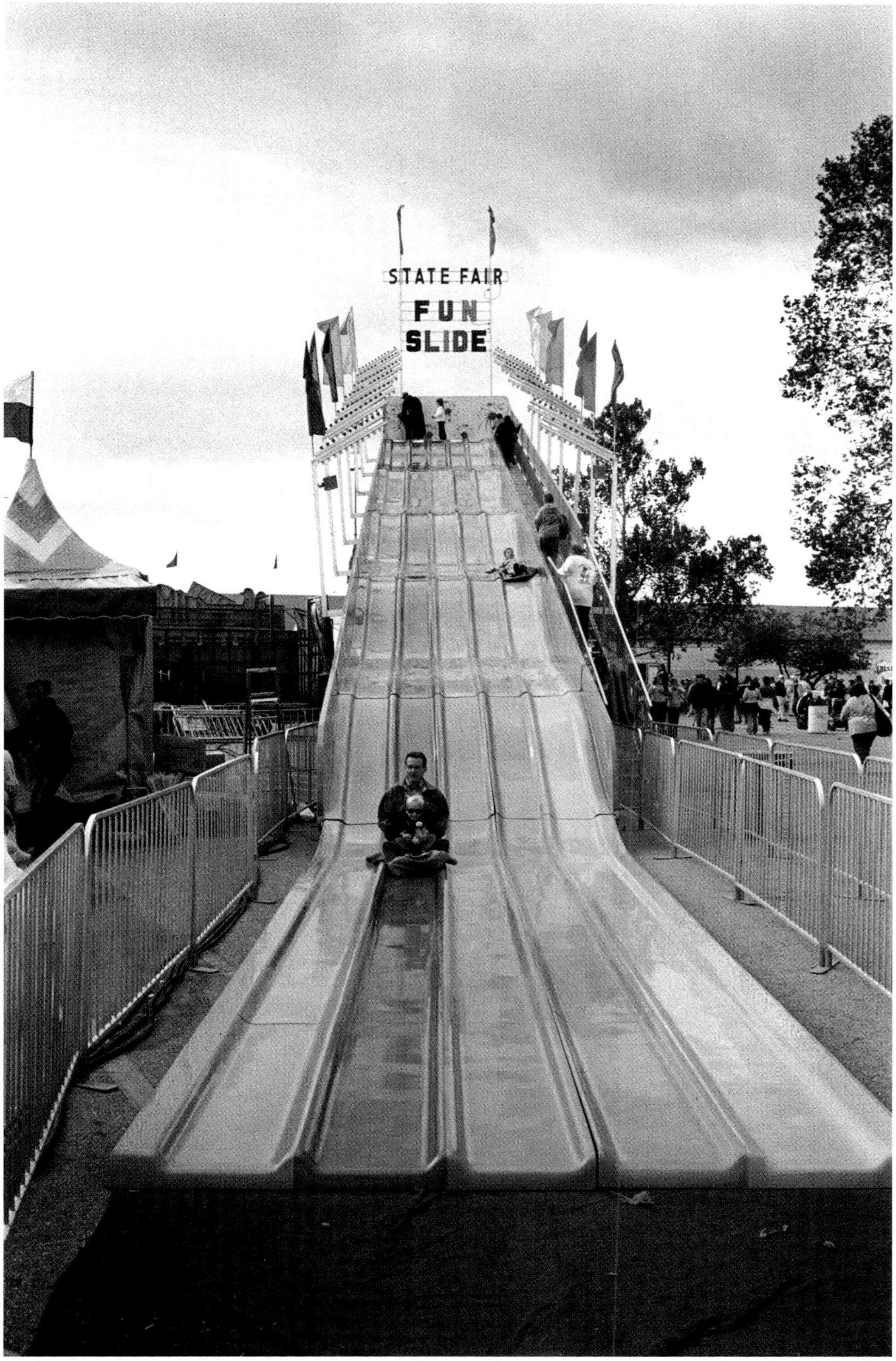
STATE FAIR
FUN
SLIDE

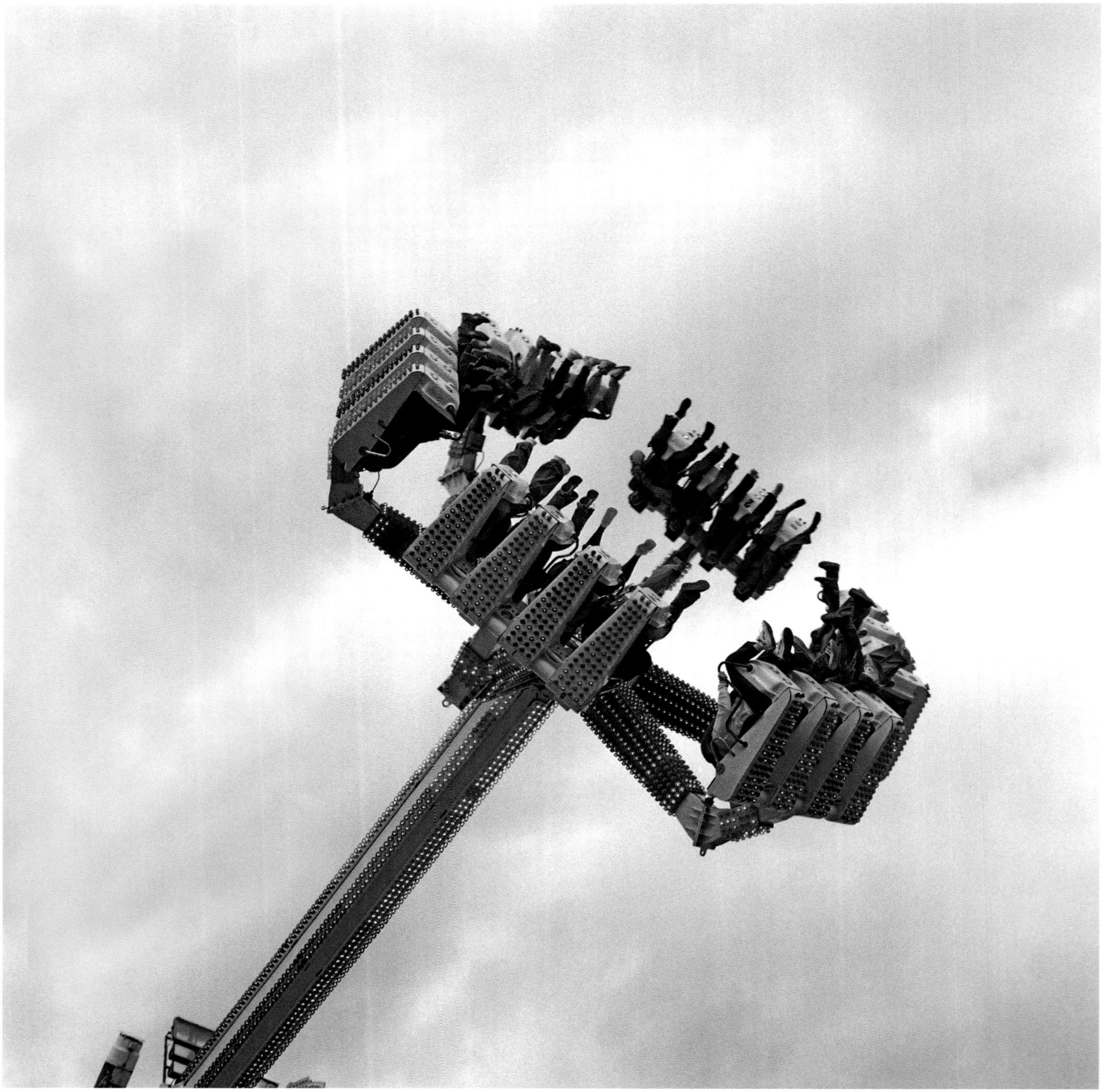

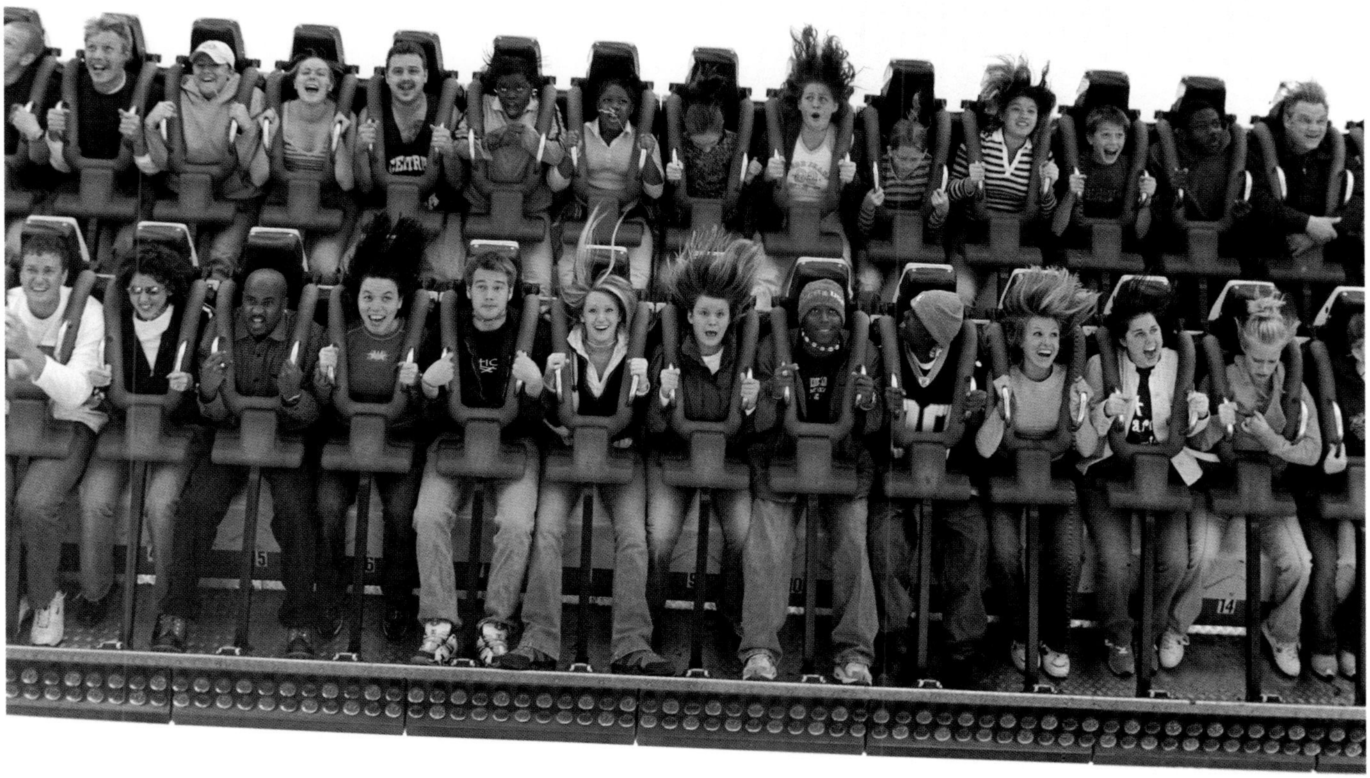

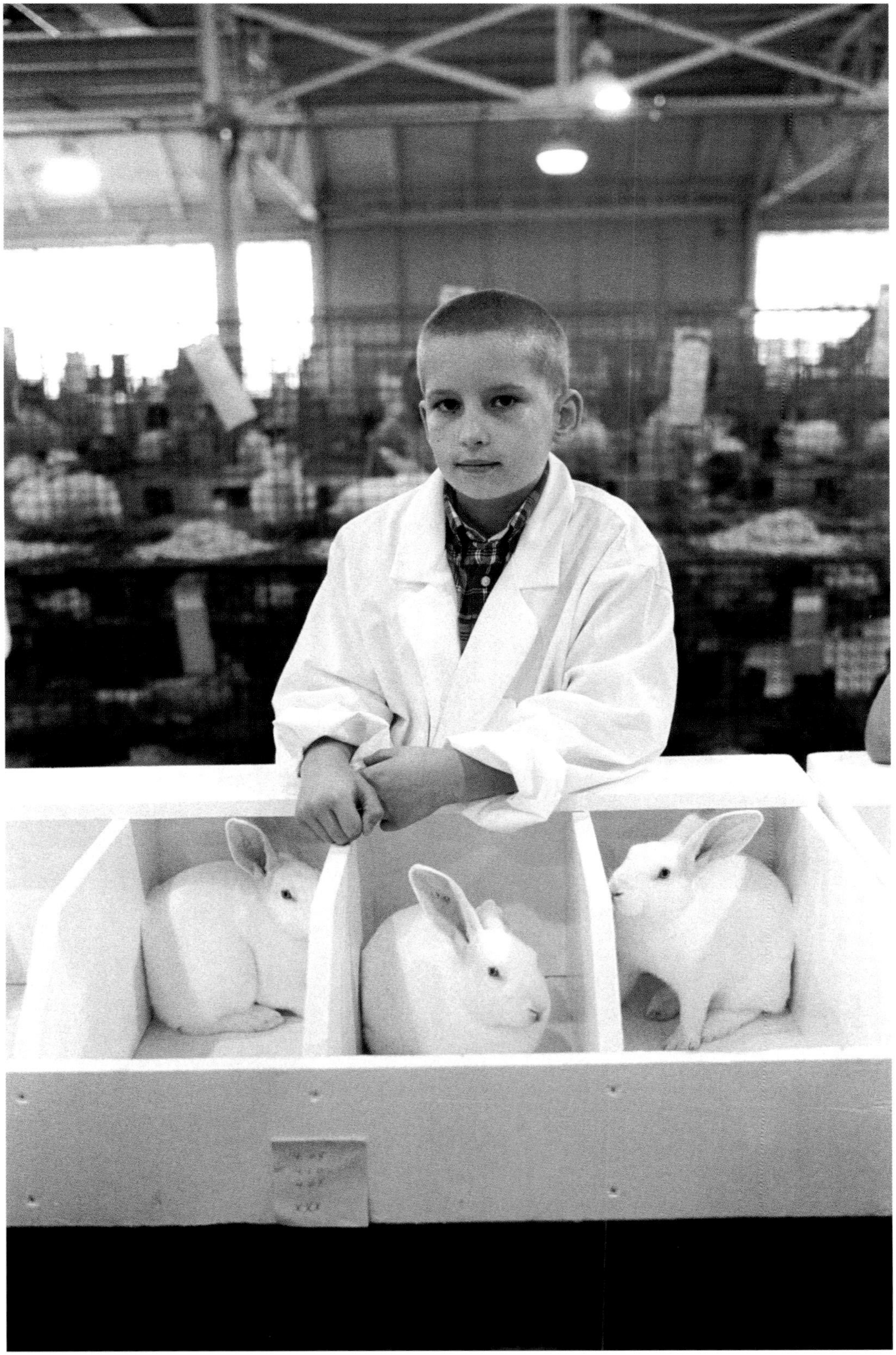

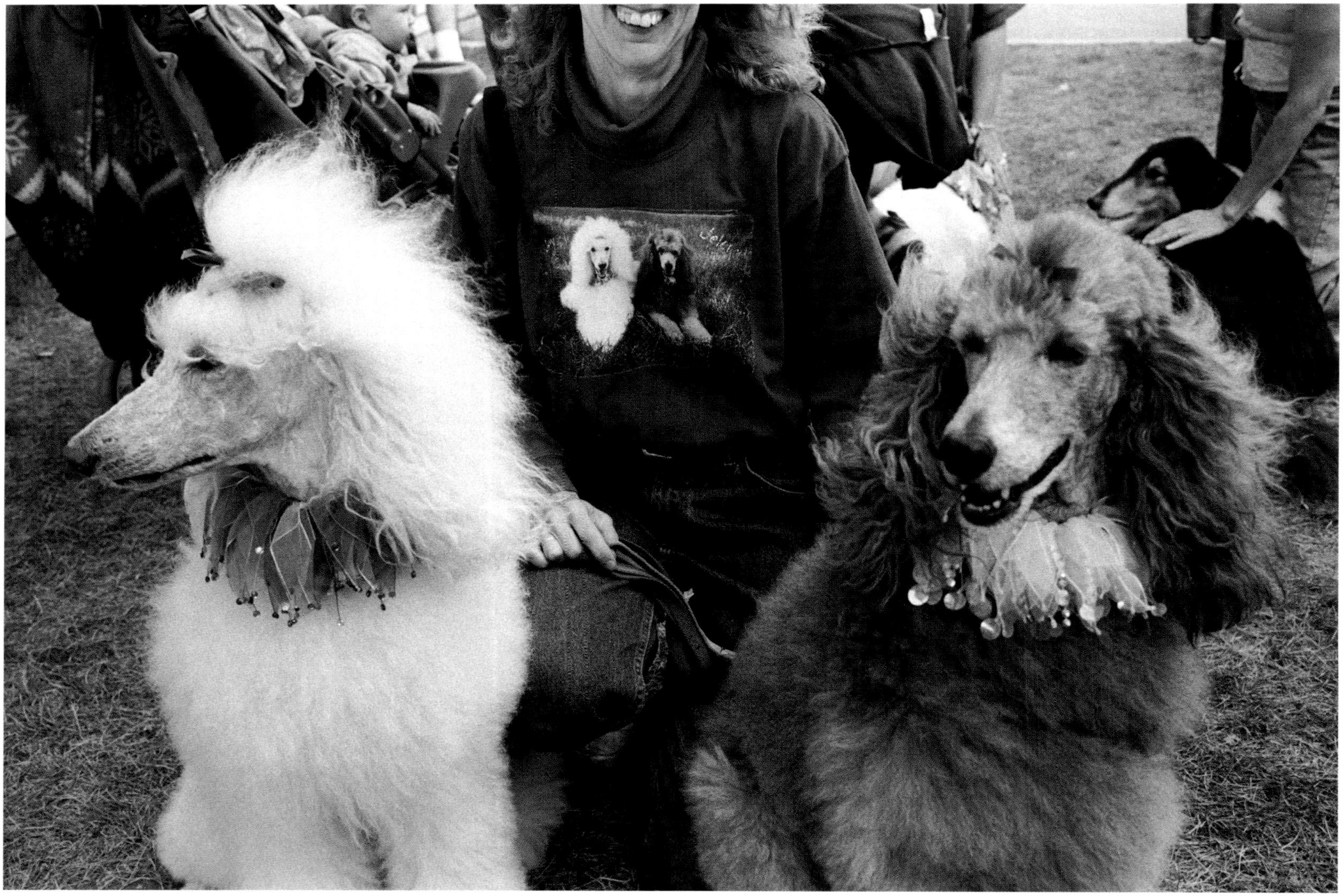

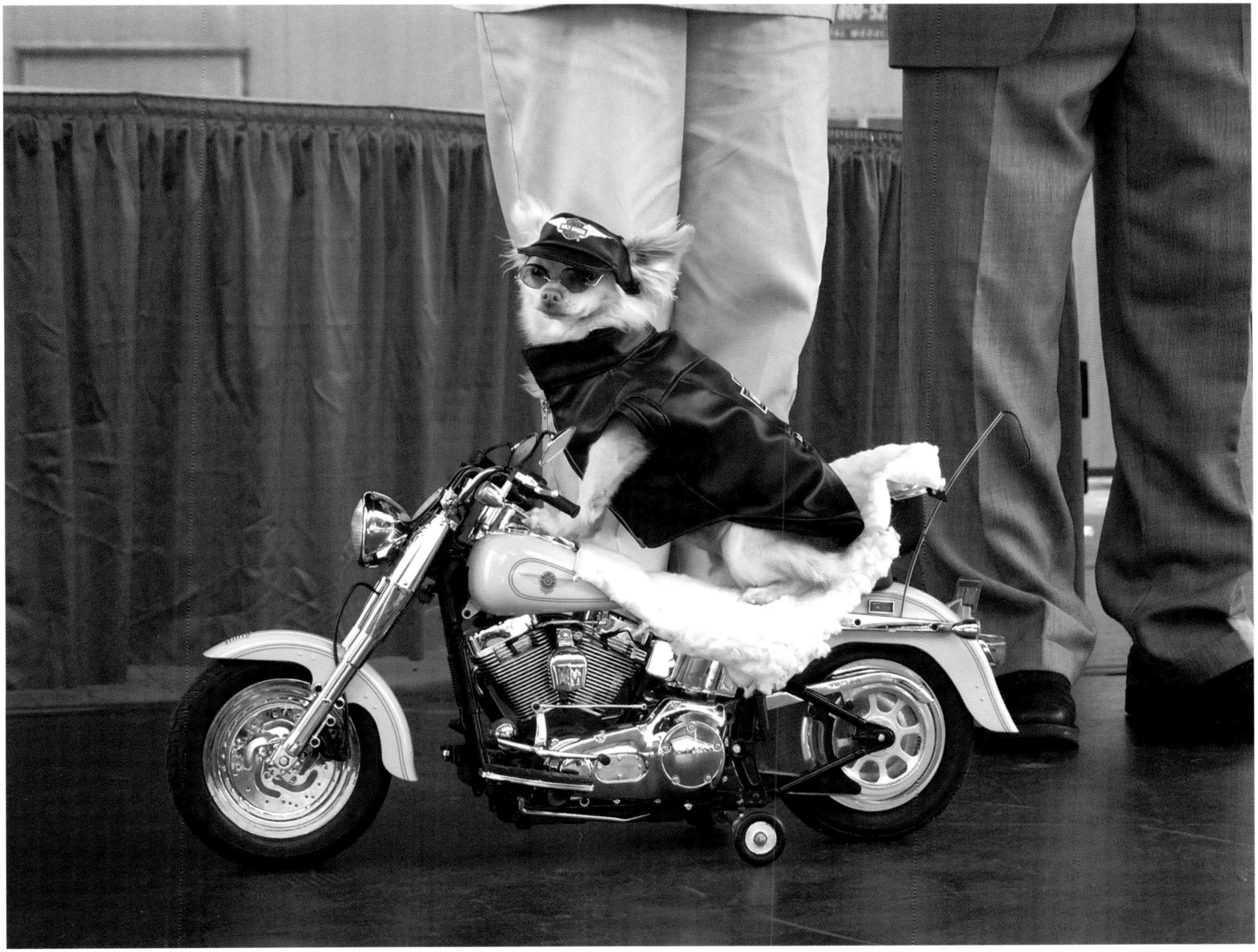

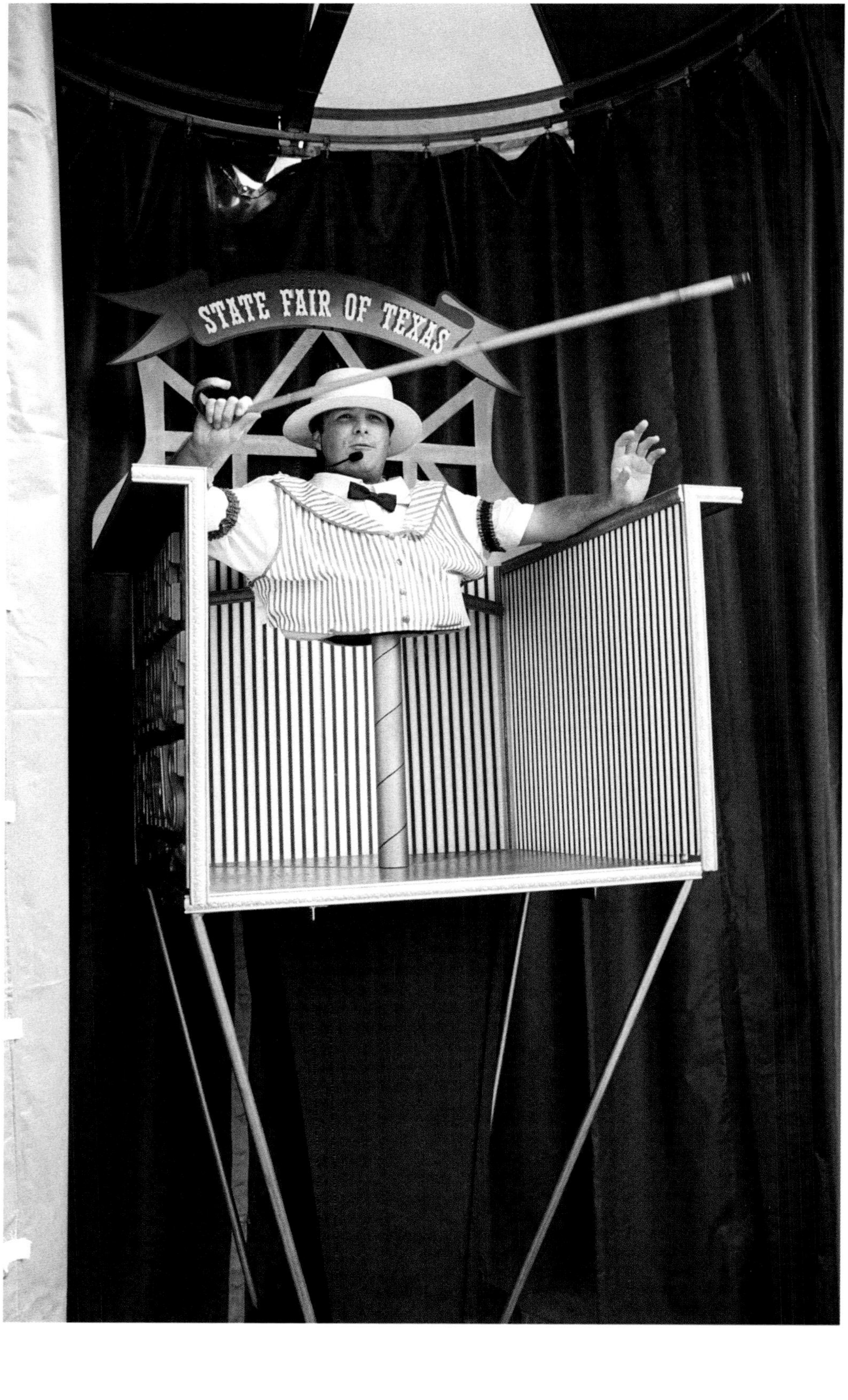
STATE FAIR OF TEXAS

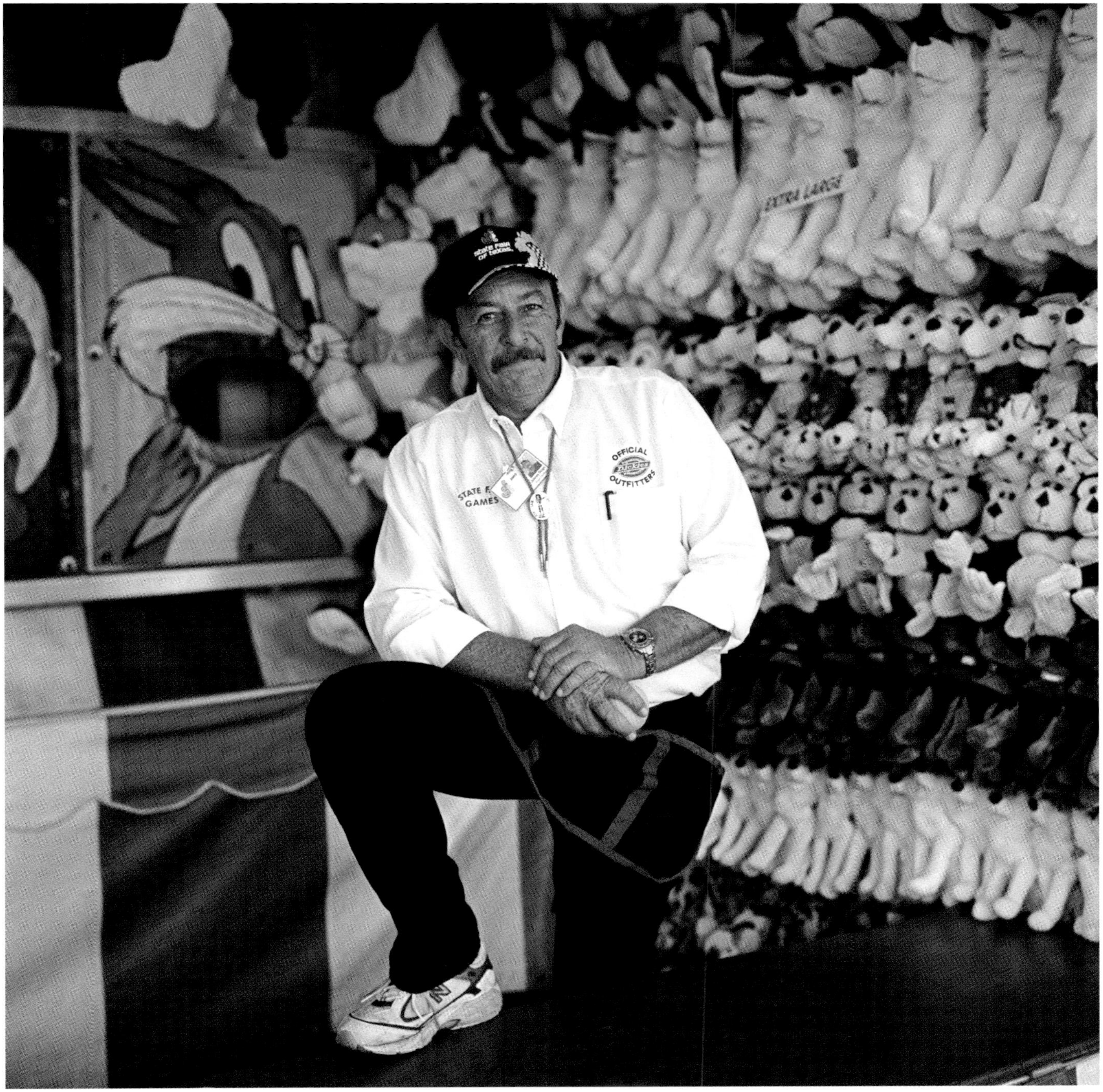
EXTRA LARGE
OFFICIAL
OUTFITTERS

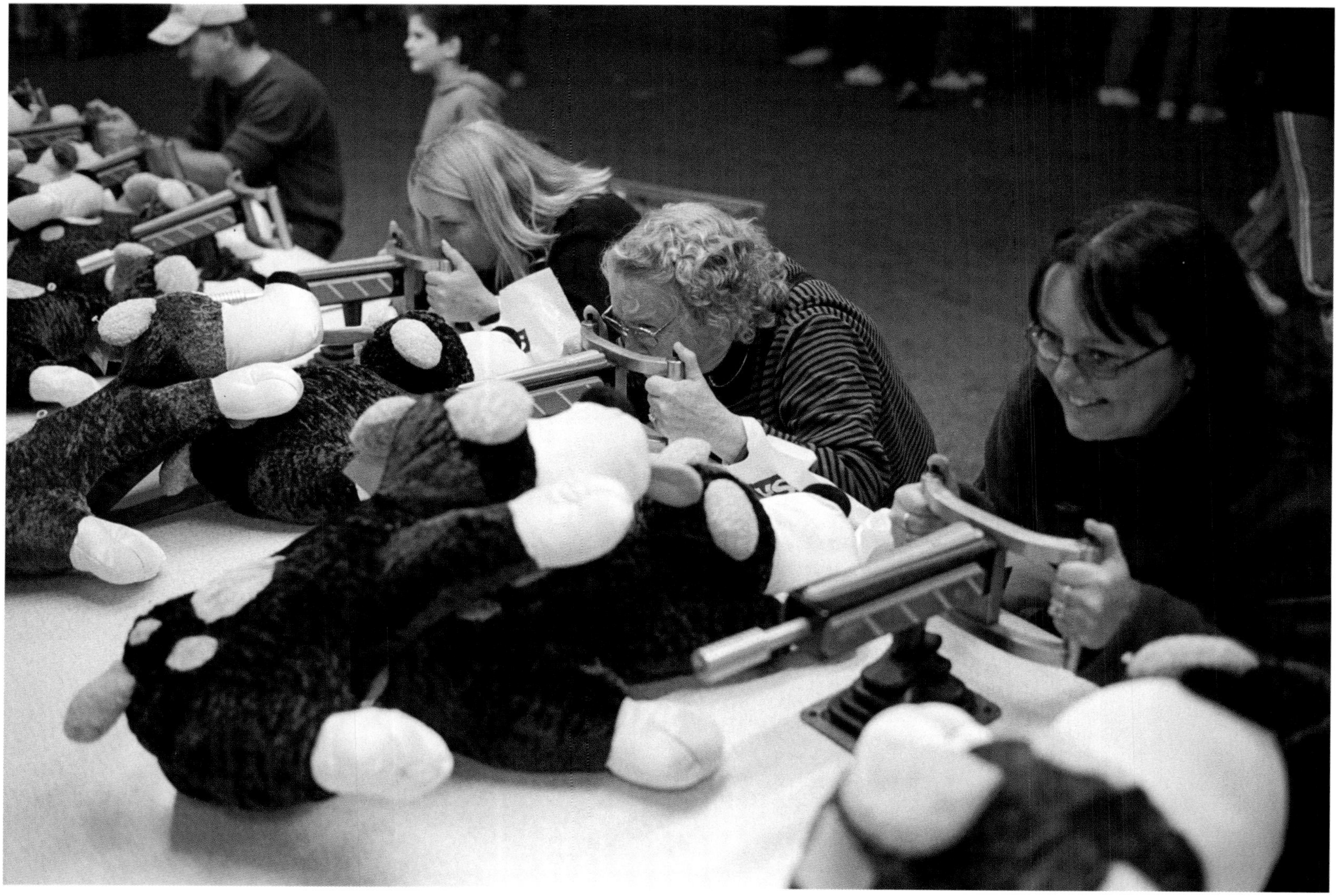

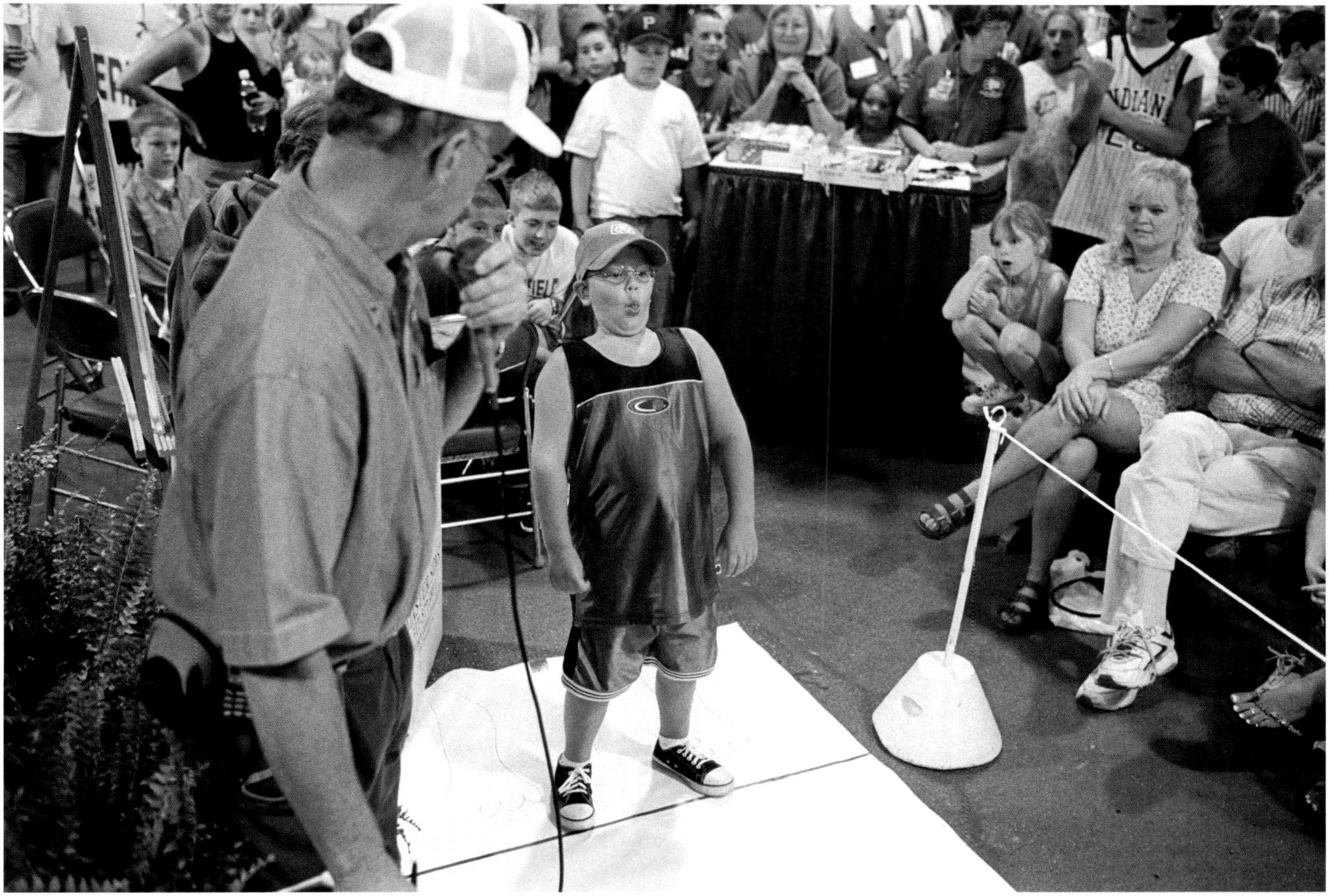
INDIANA

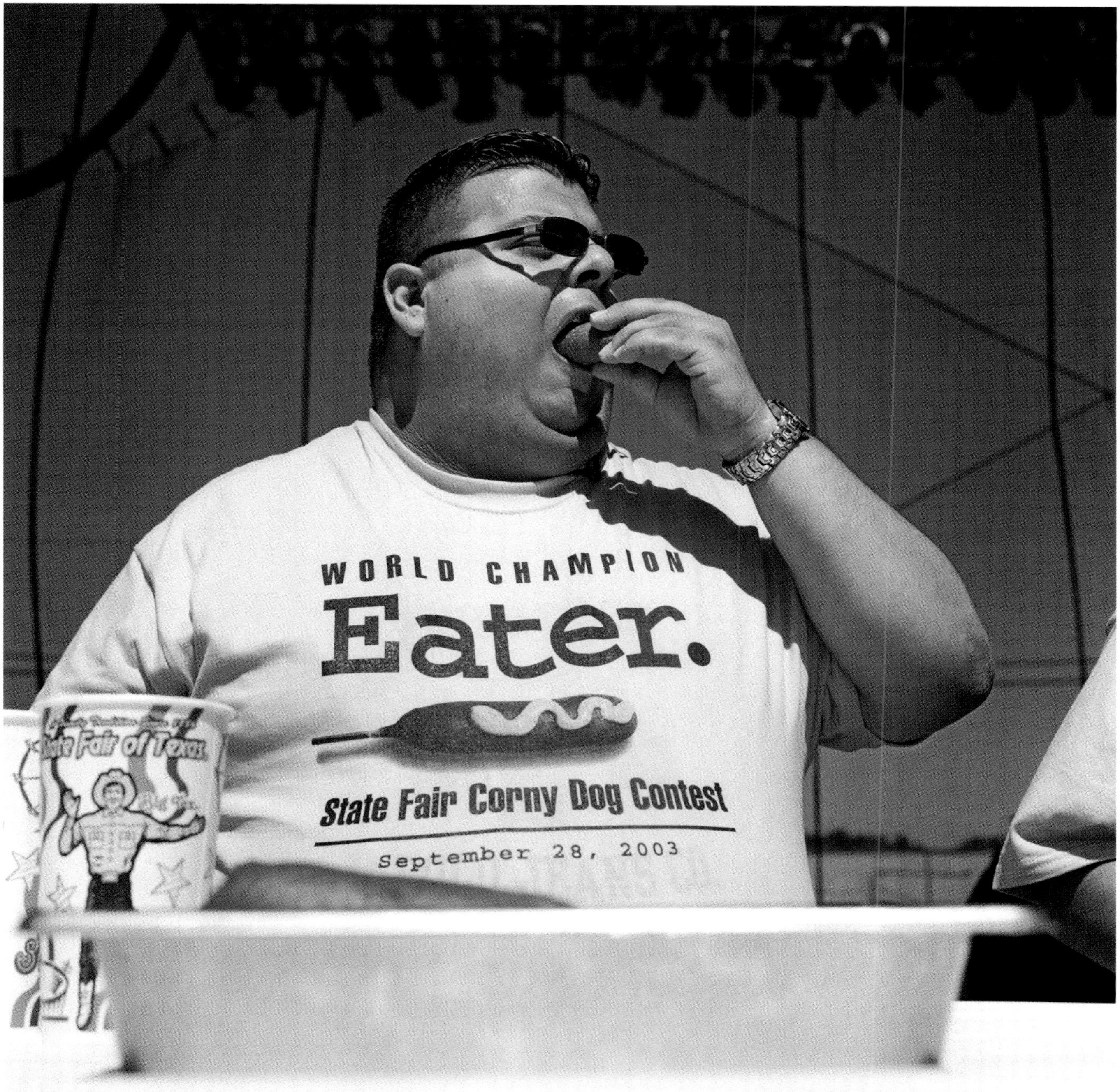
WORLD CHAMPION
Eater.
State Fair Corny Dog Contest
September 28, 2003
State Fair of Texas

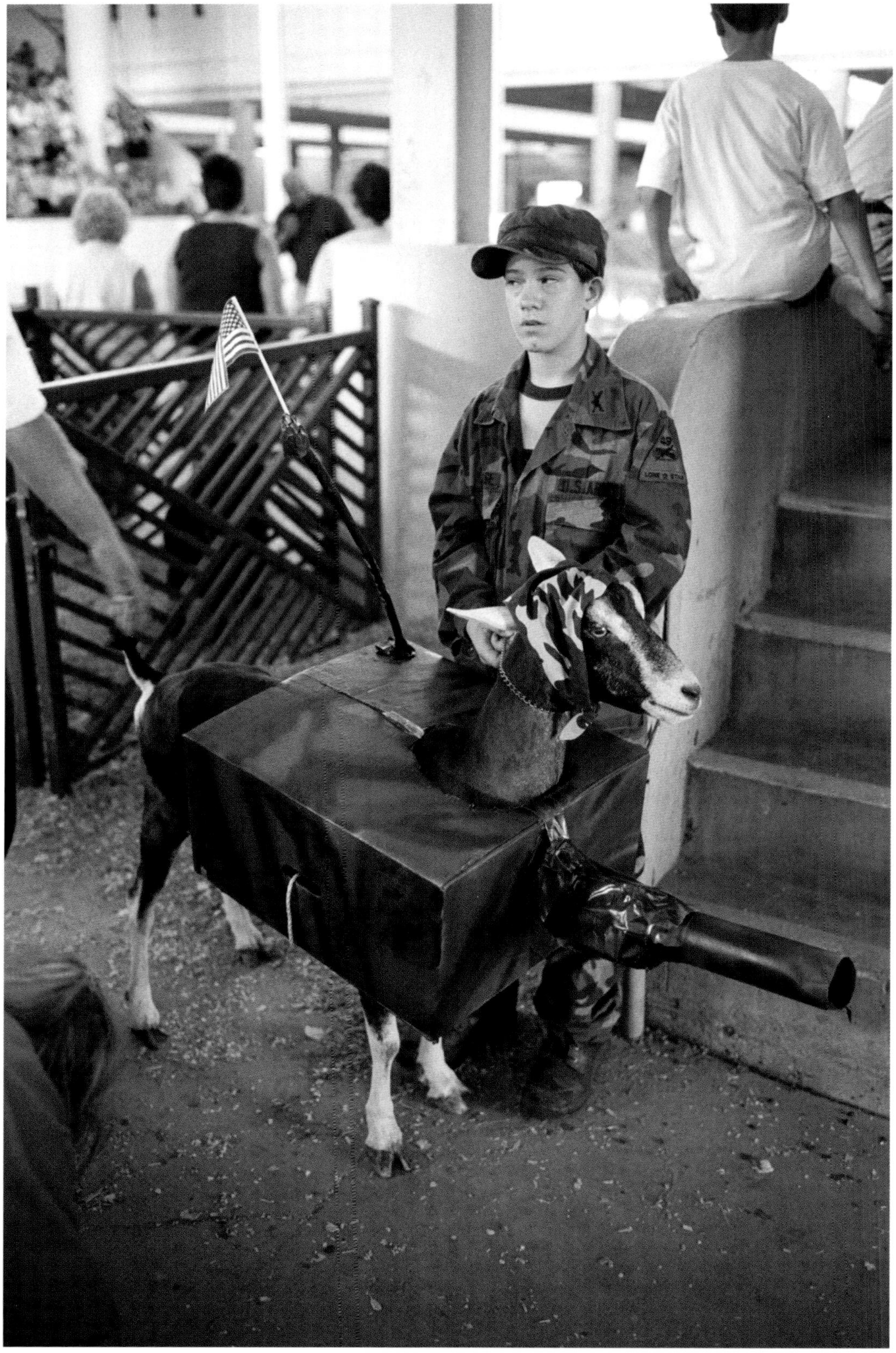

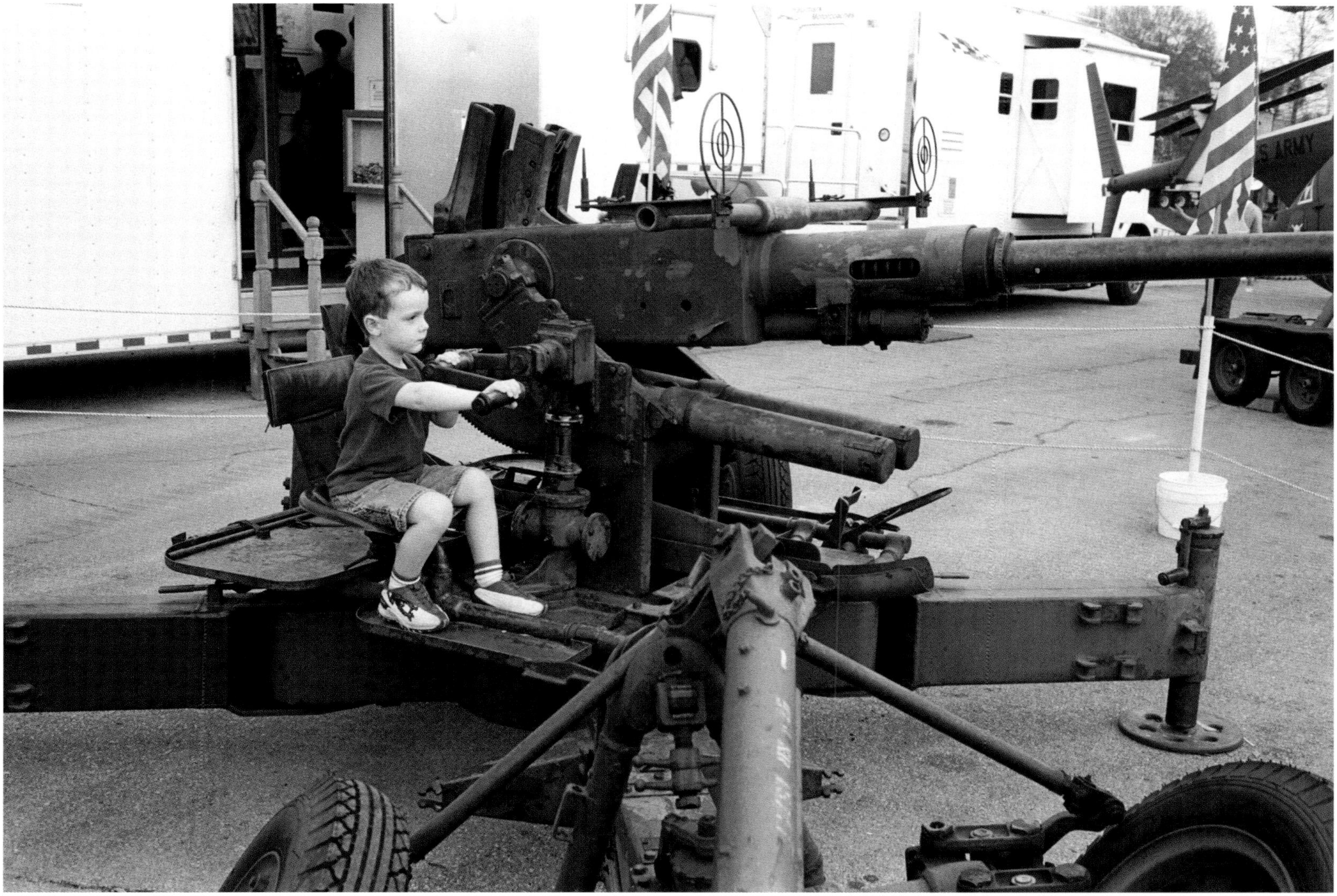

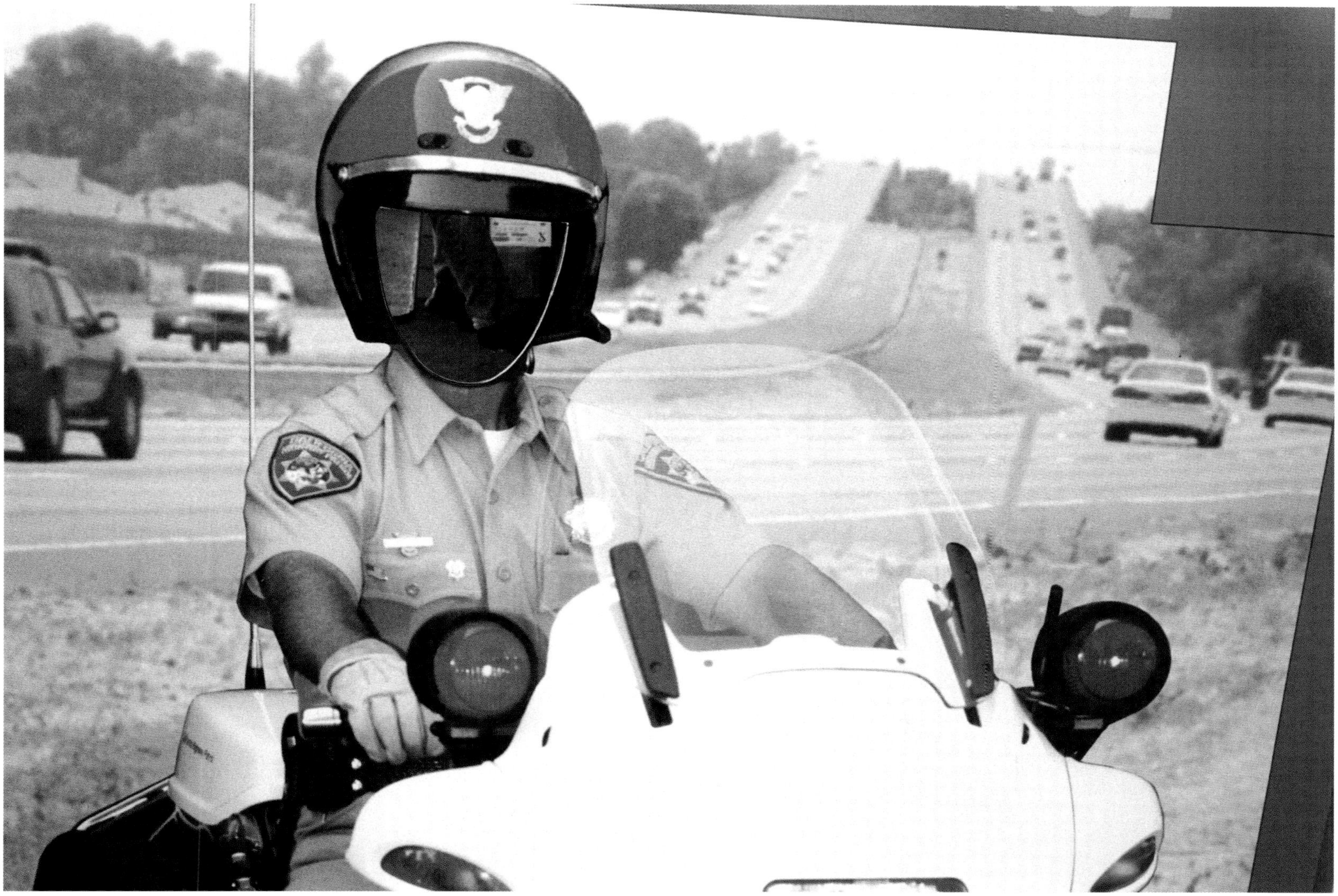

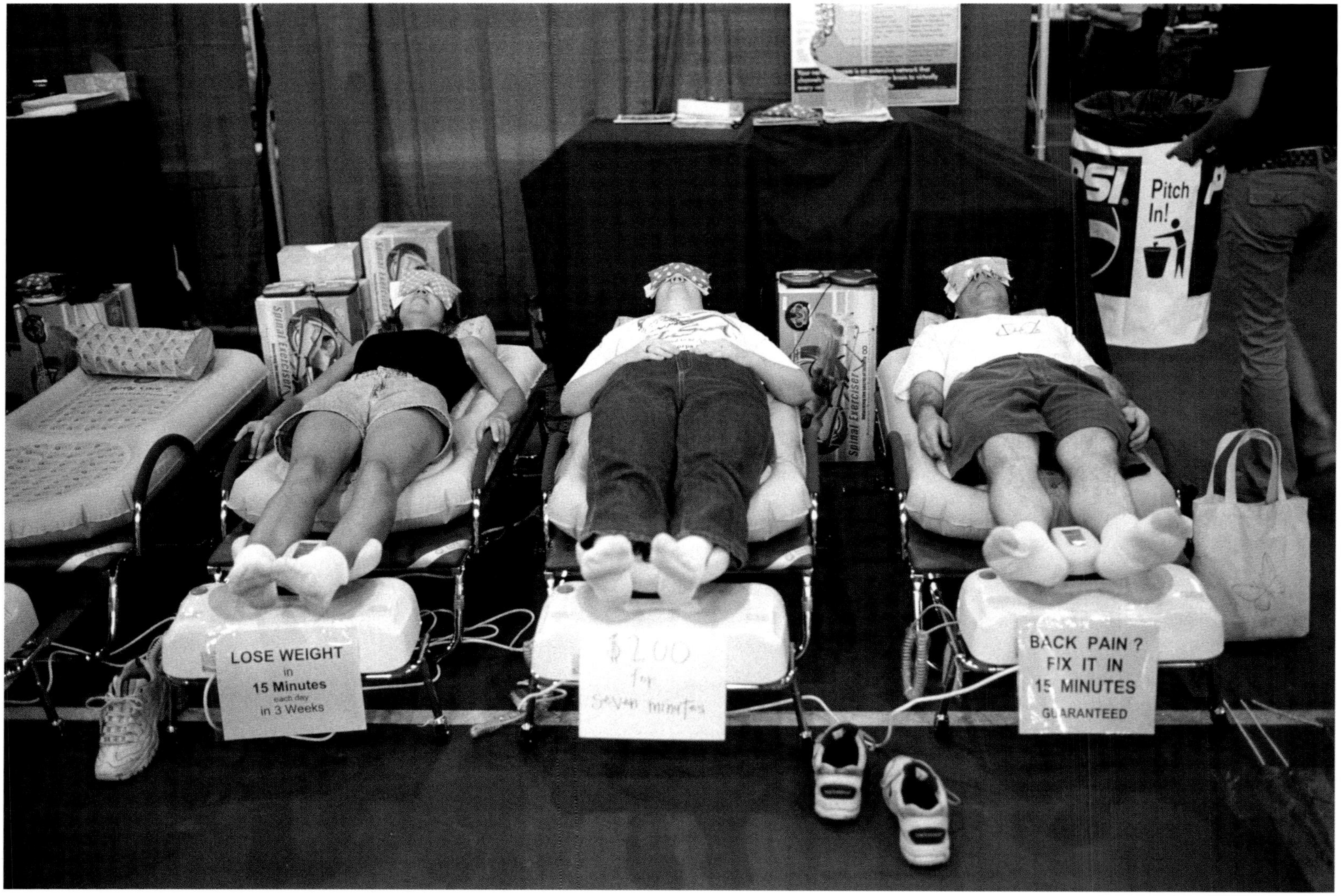
Spinal Exerciser
LOSE WEIGHT
in
15 Minutes
each day
in 3 Weeks
BACK PAIN ?
FIX IT IN
15 MINUTES
GUARANTEED
Pitch
In!

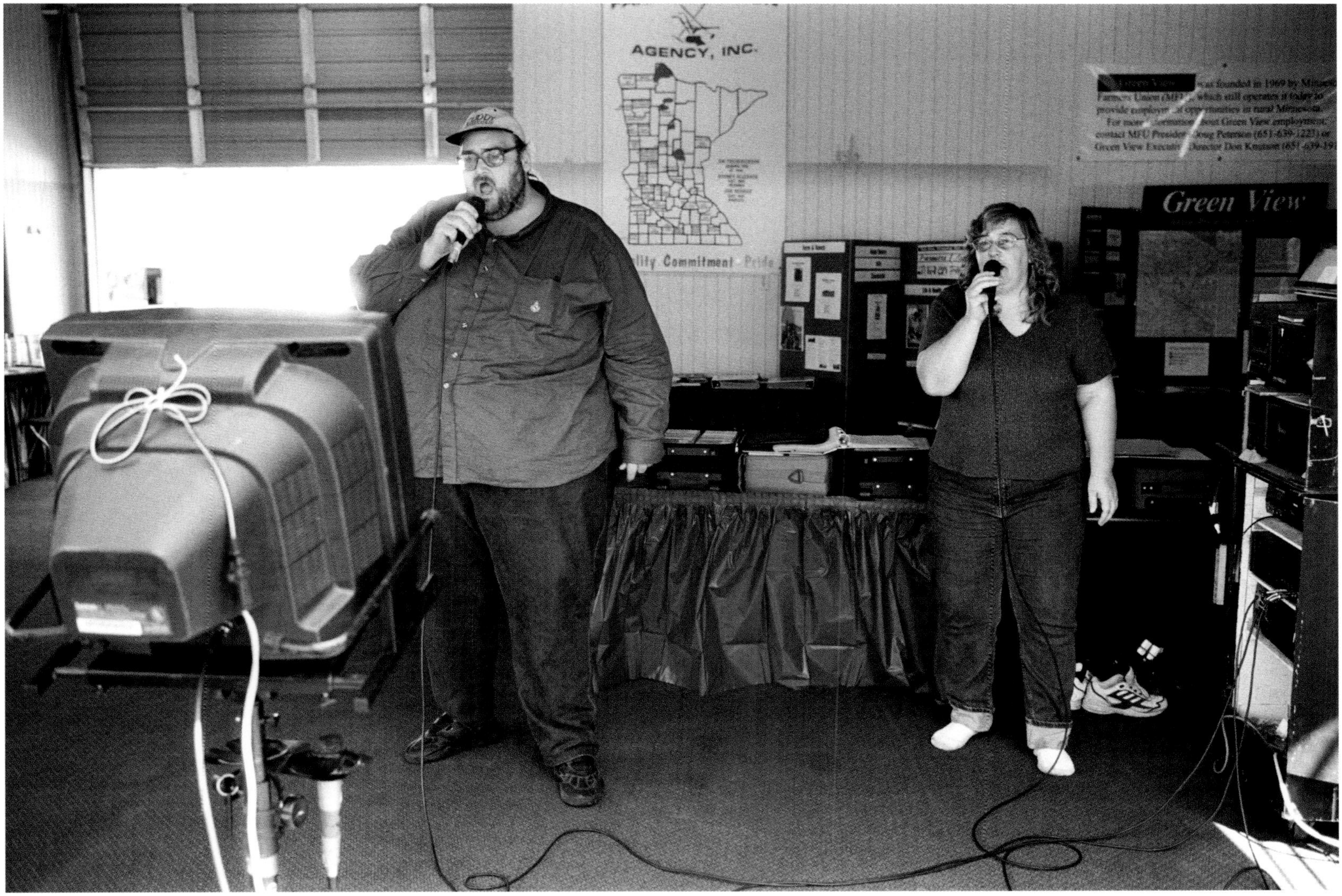
AGENCY, INC.
Green View

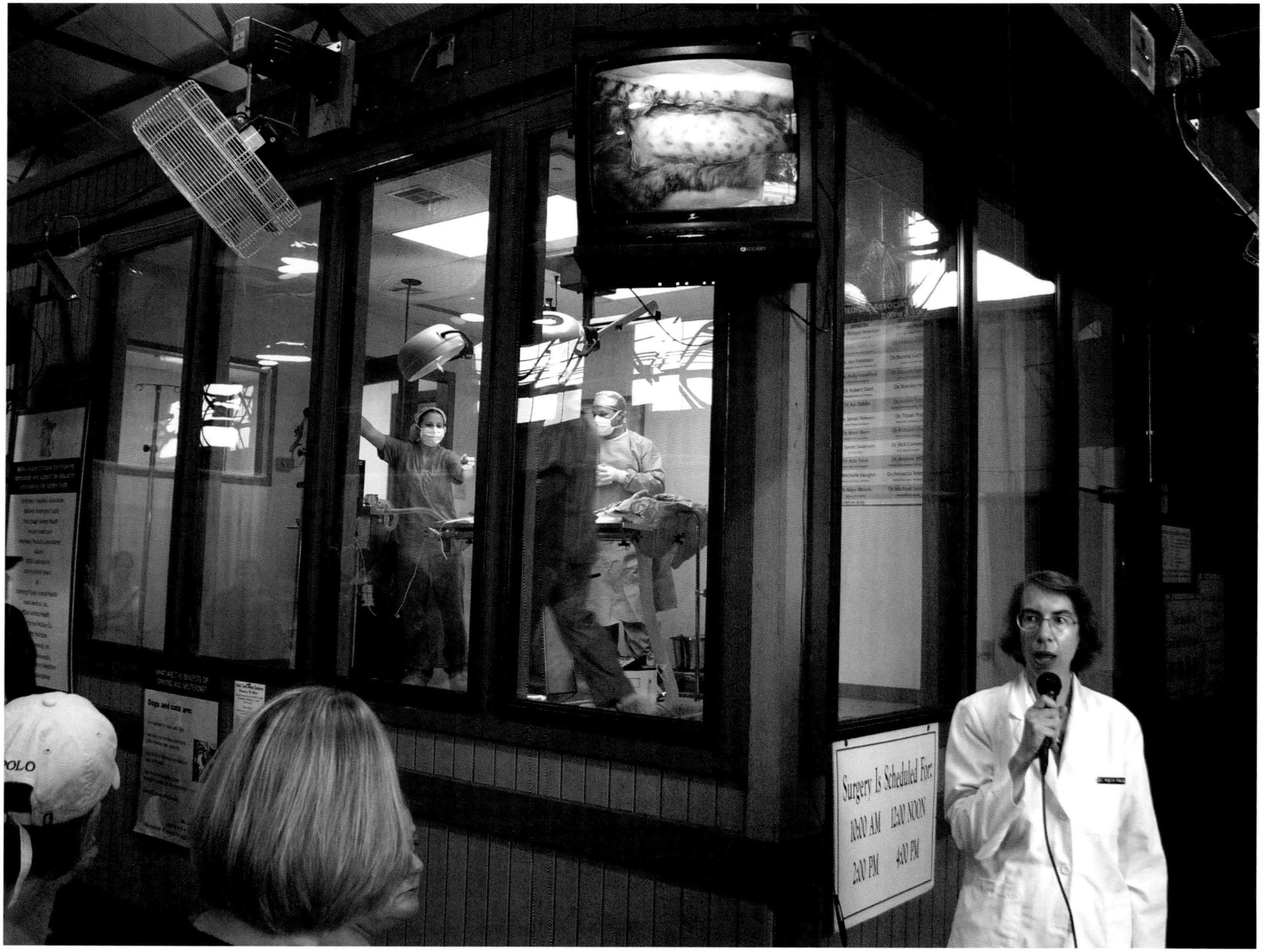
Surgery Is Scheduled For:
10:00 AM 12:00 NOON
2:00 PM 4:00 PM

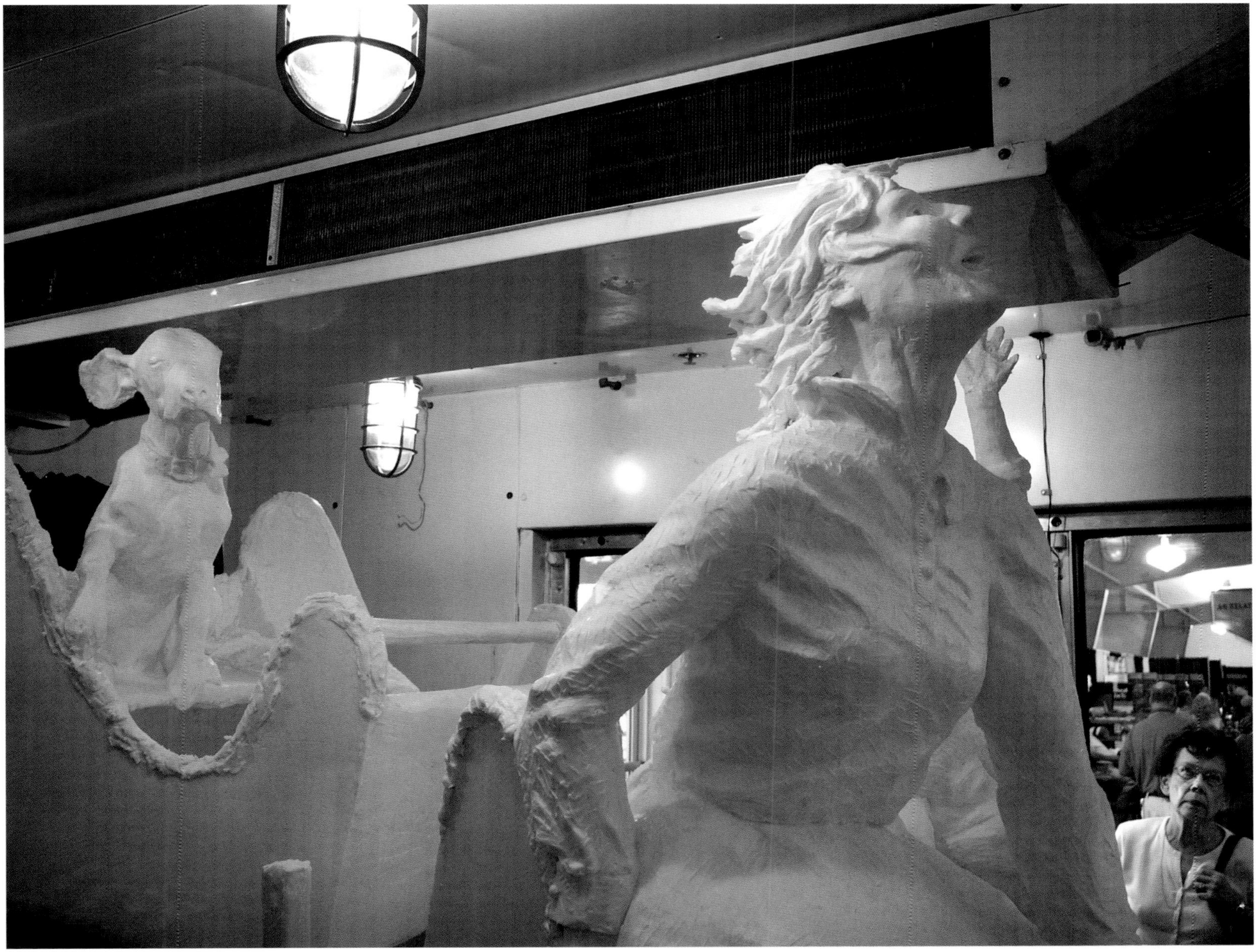

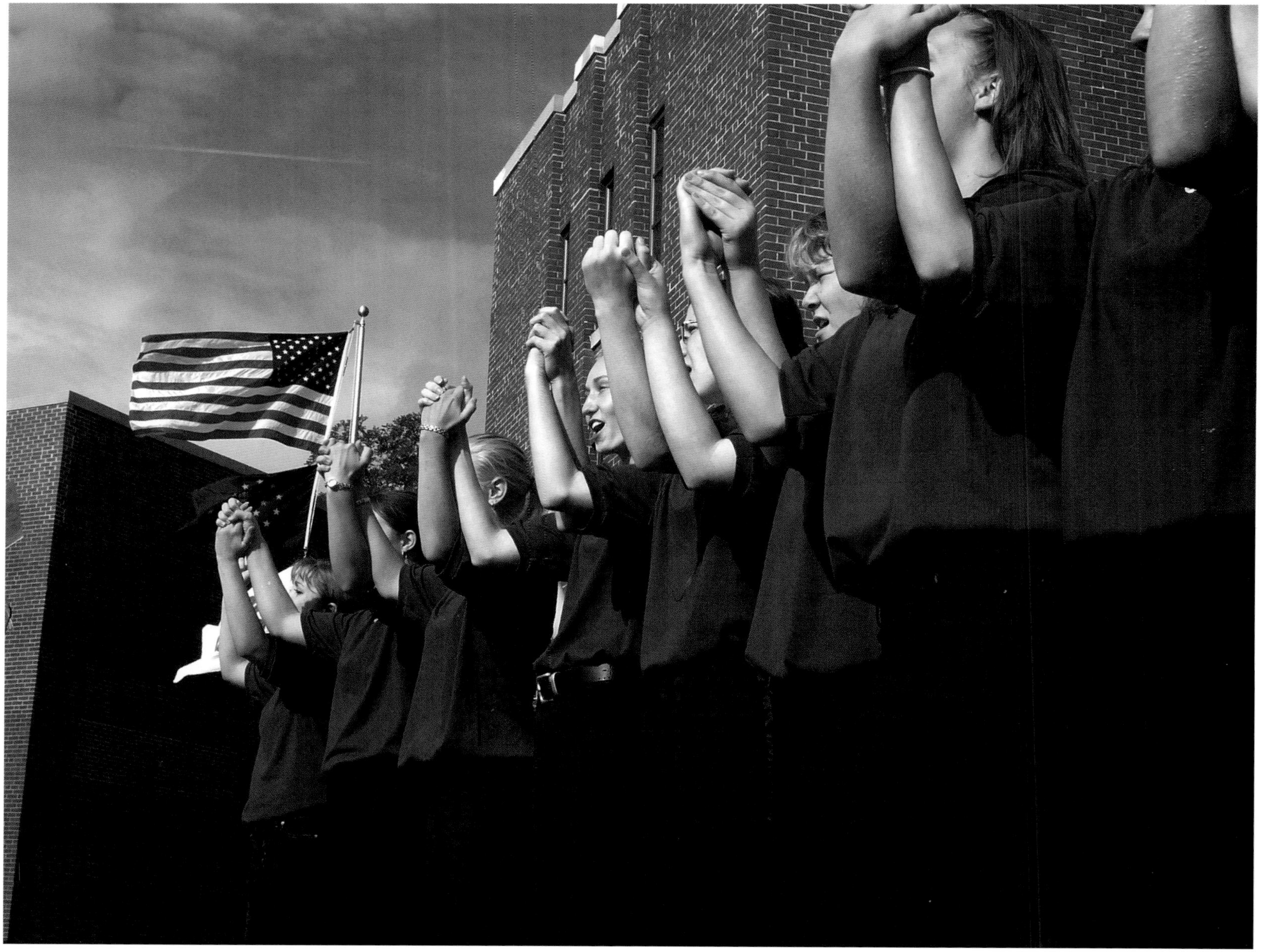

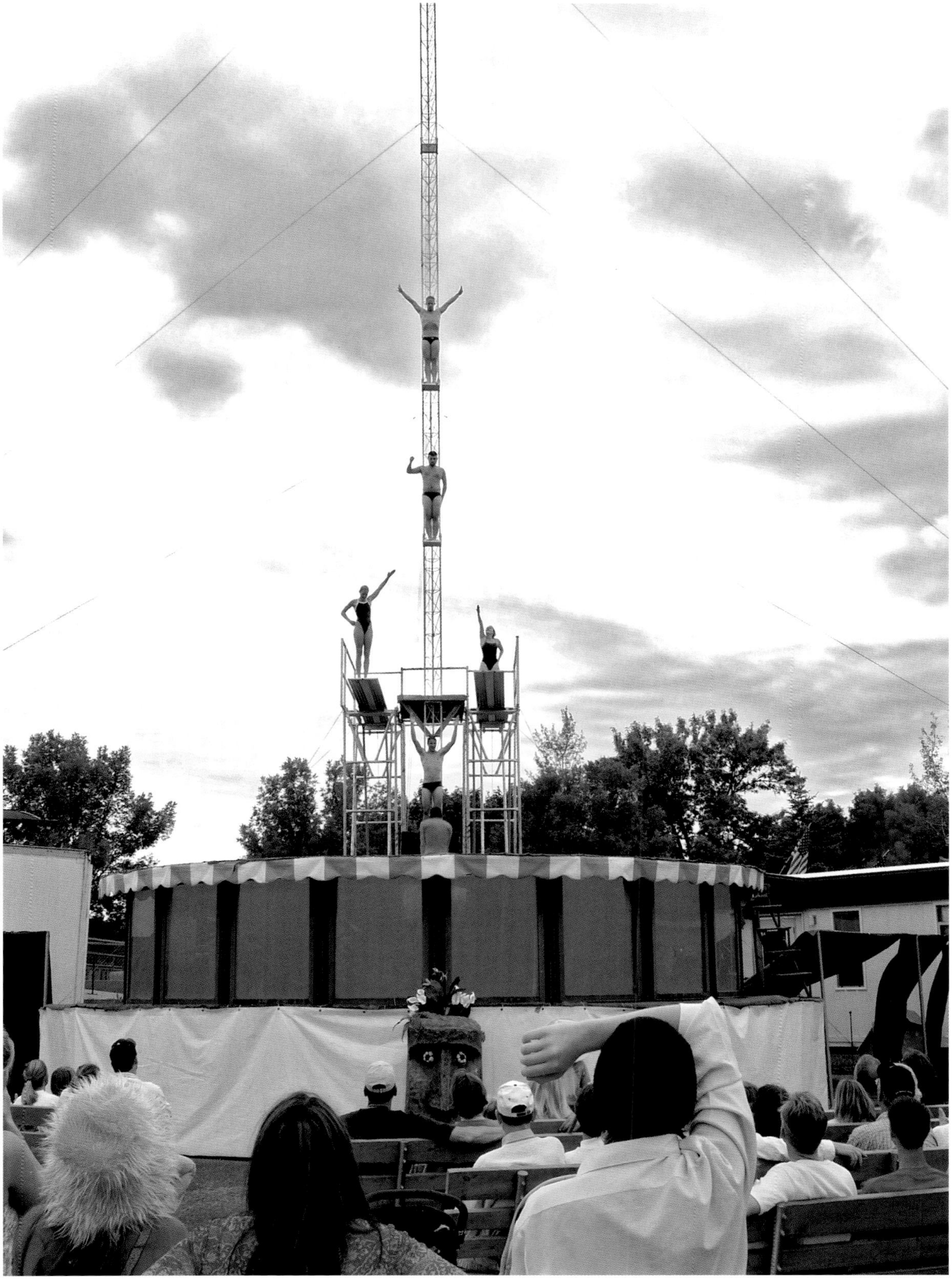

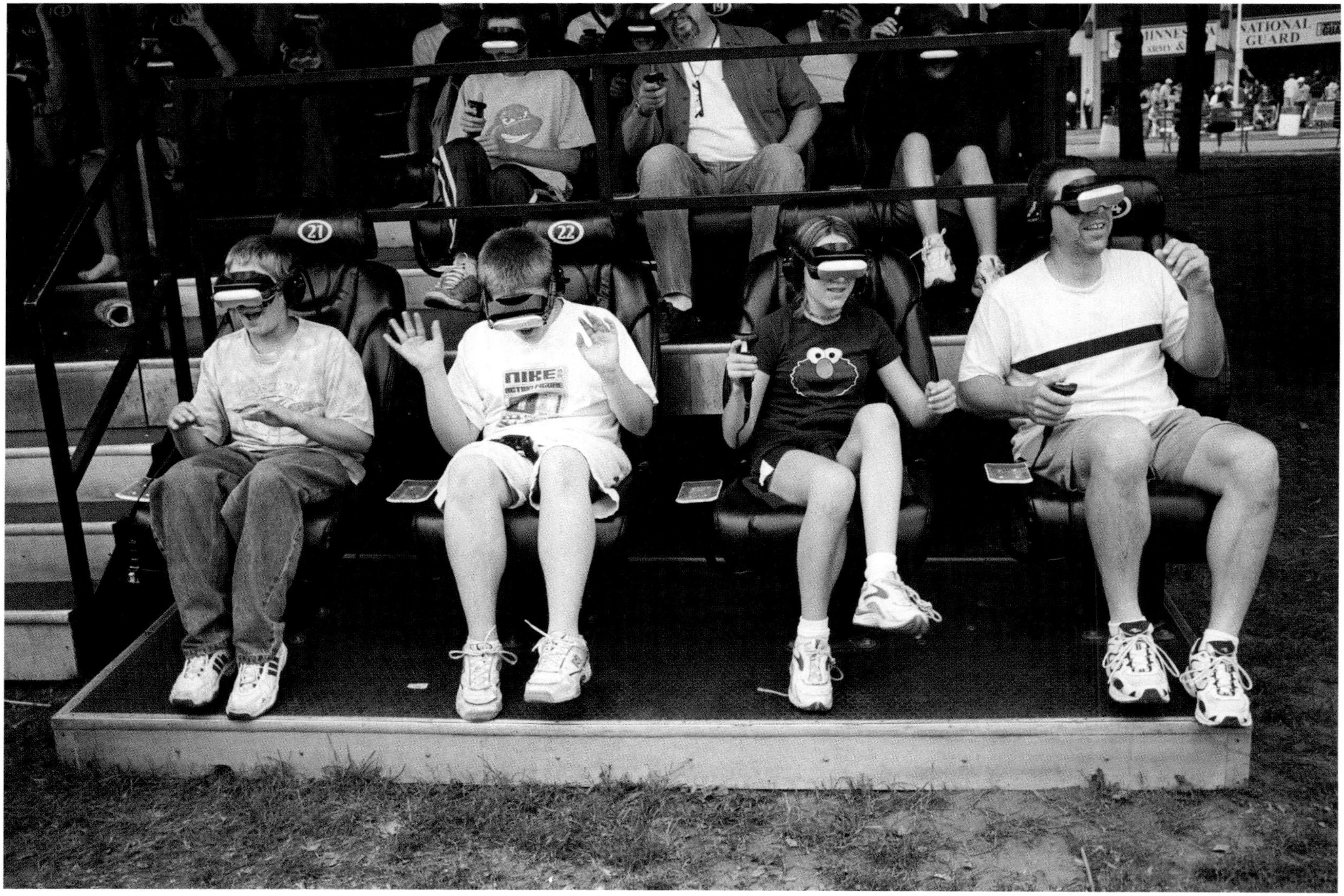
NATIONAL
GUARD
21
22
NIKE

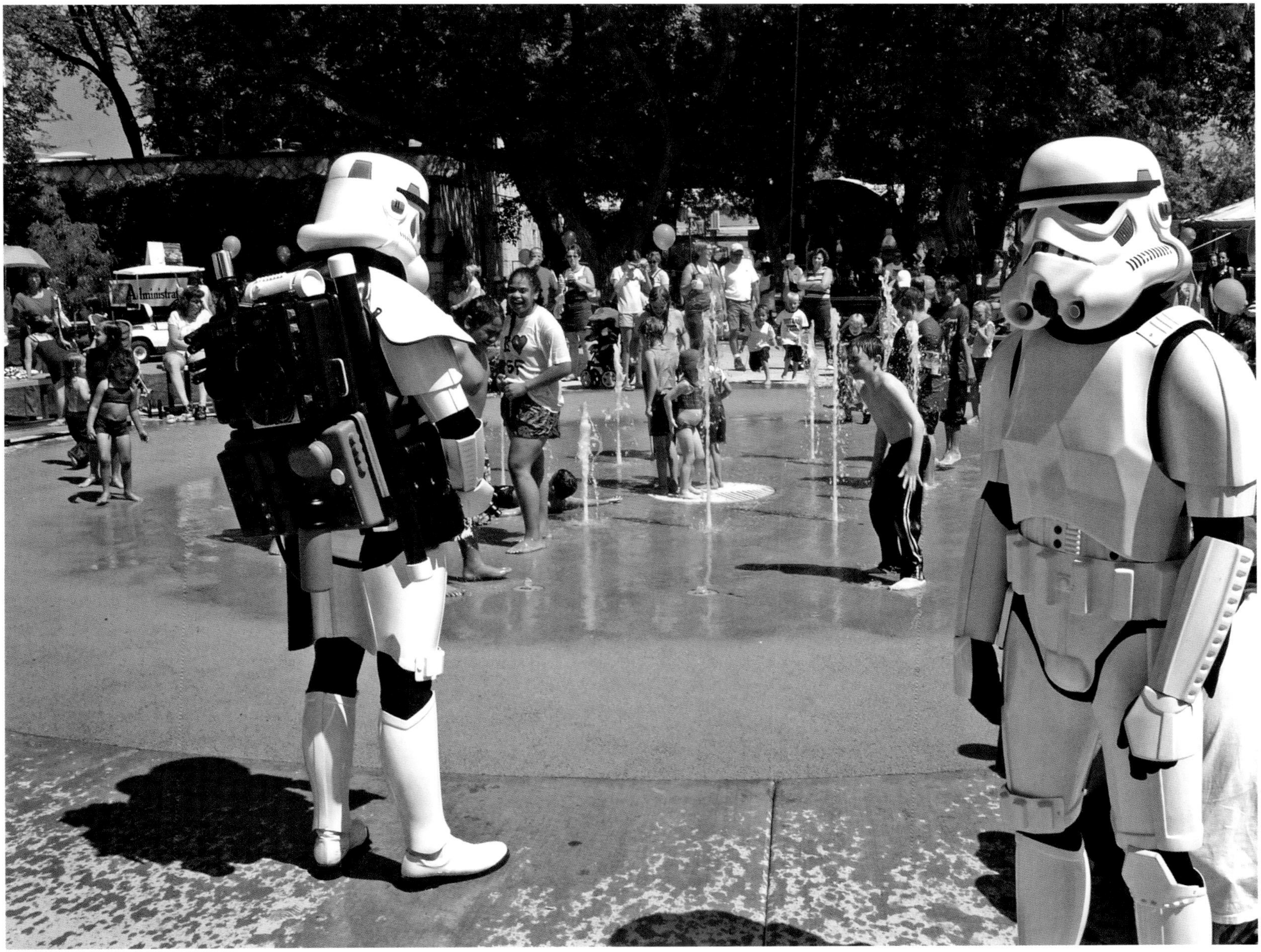

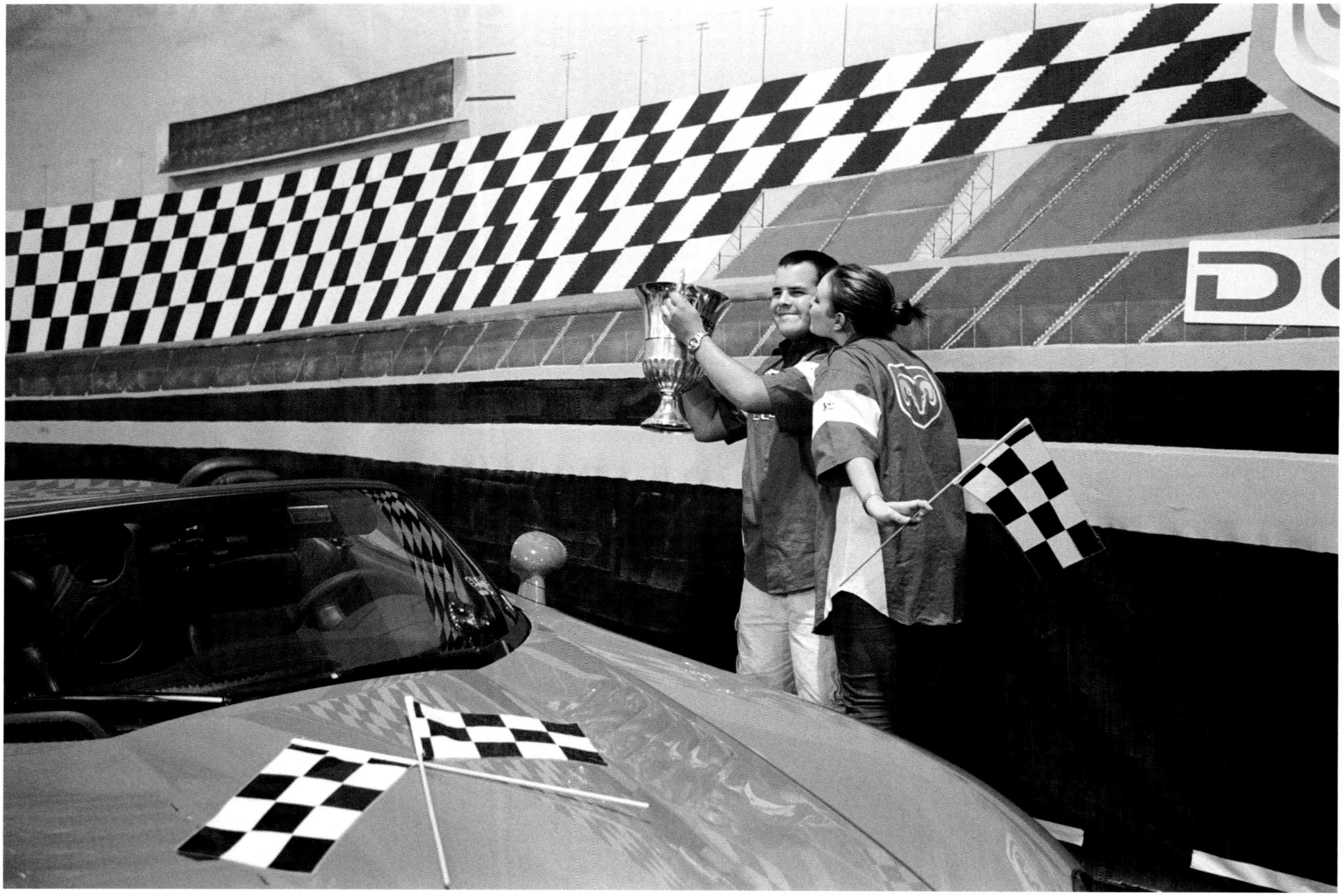

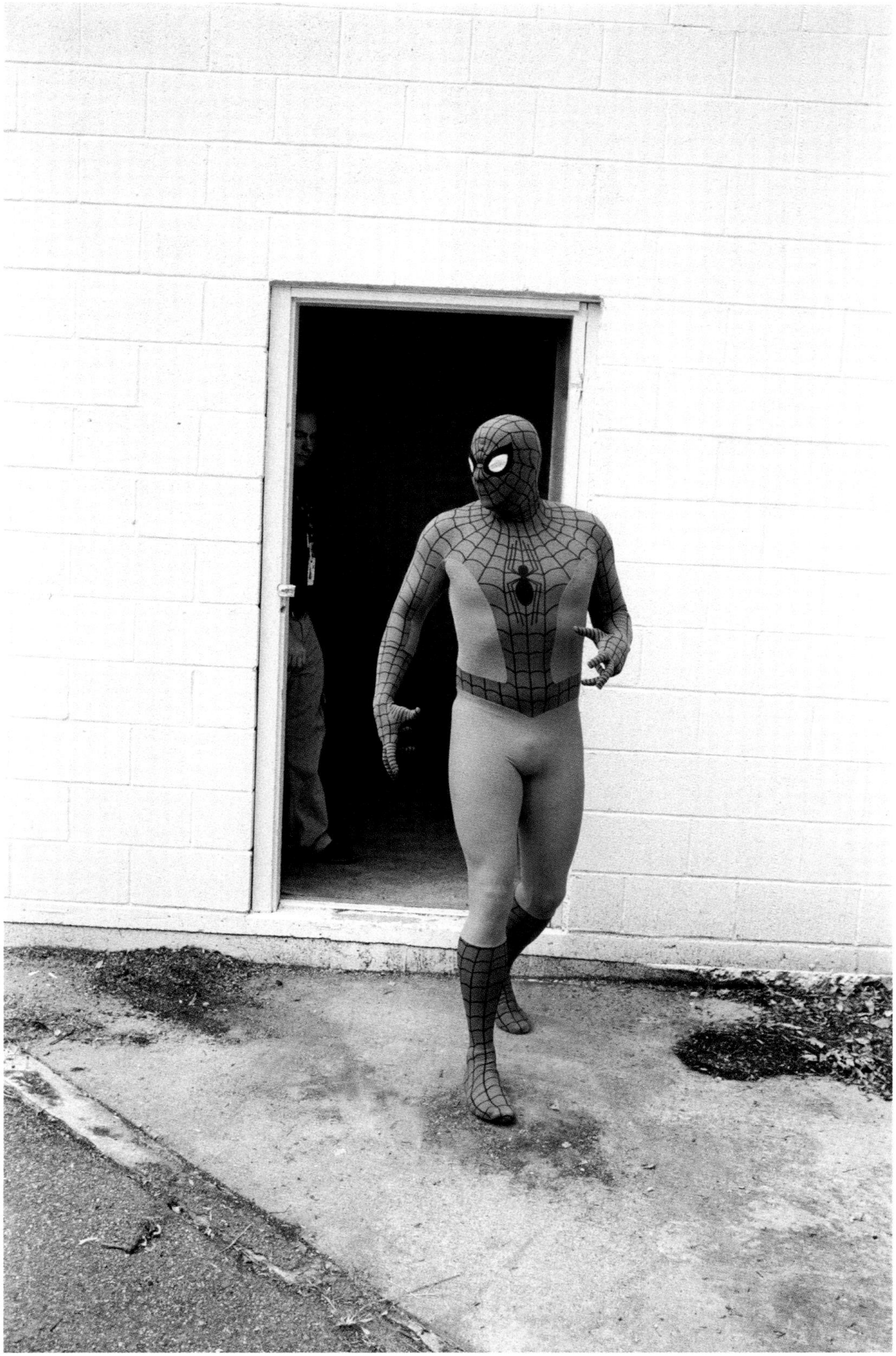

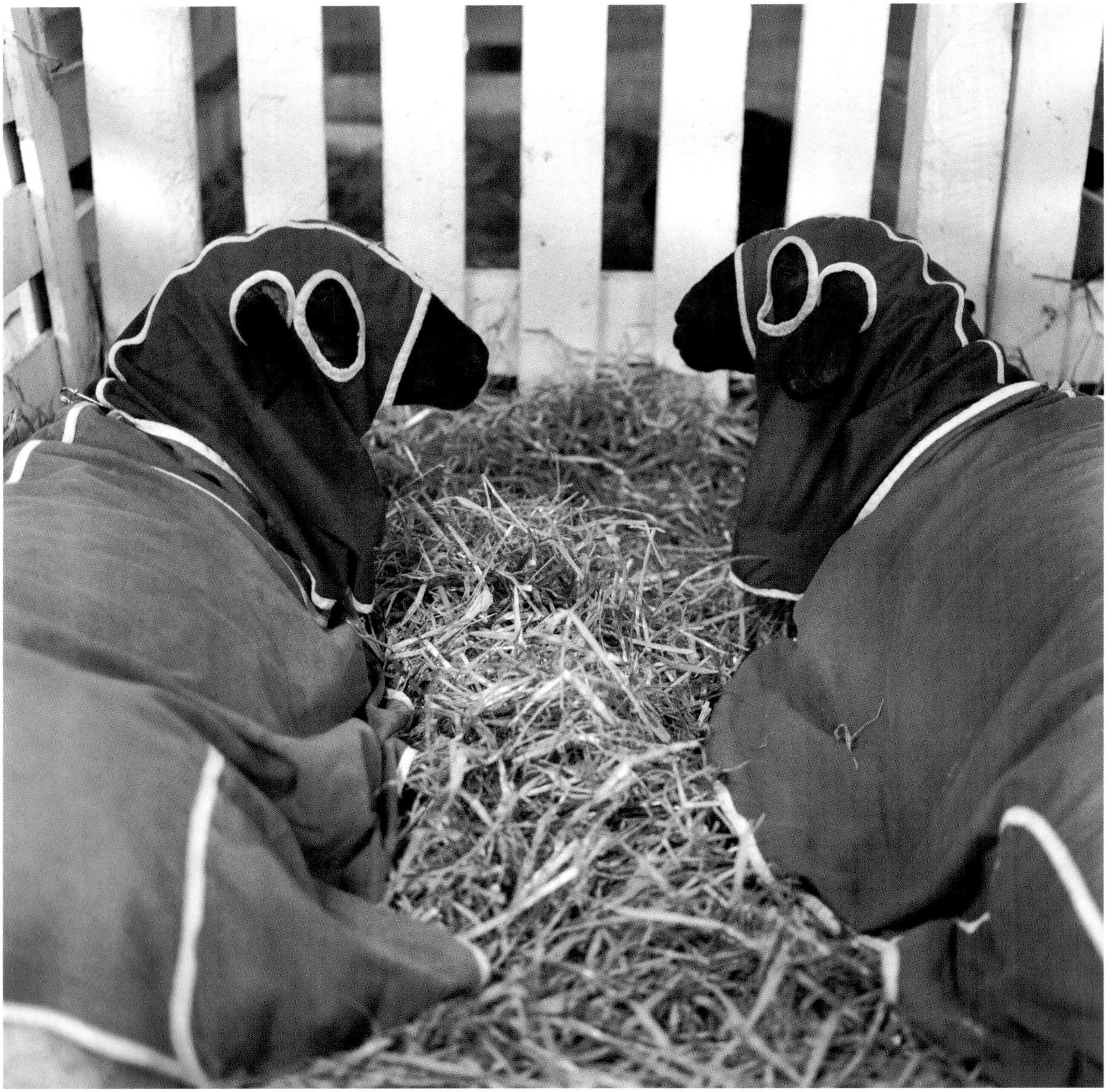

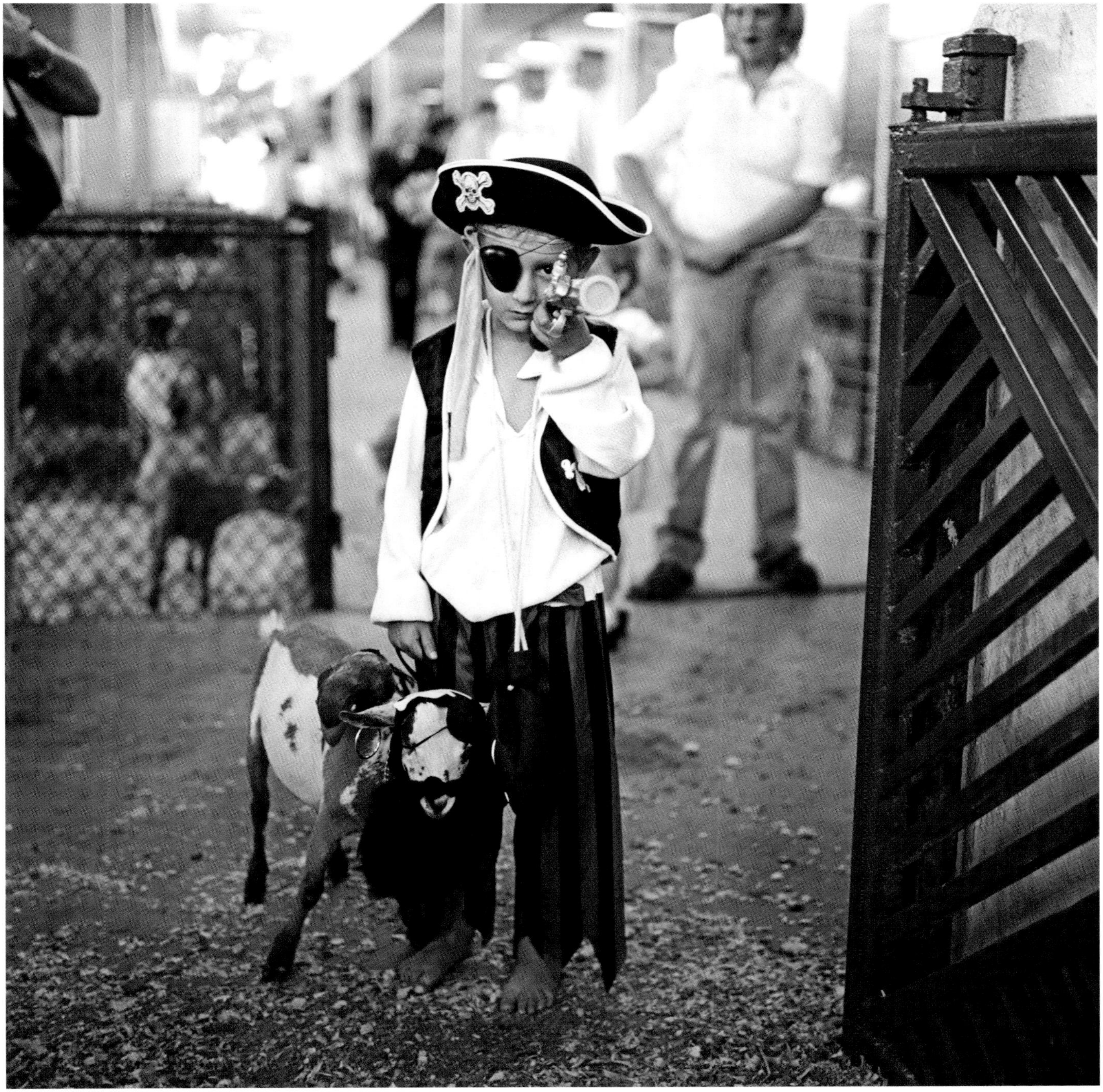

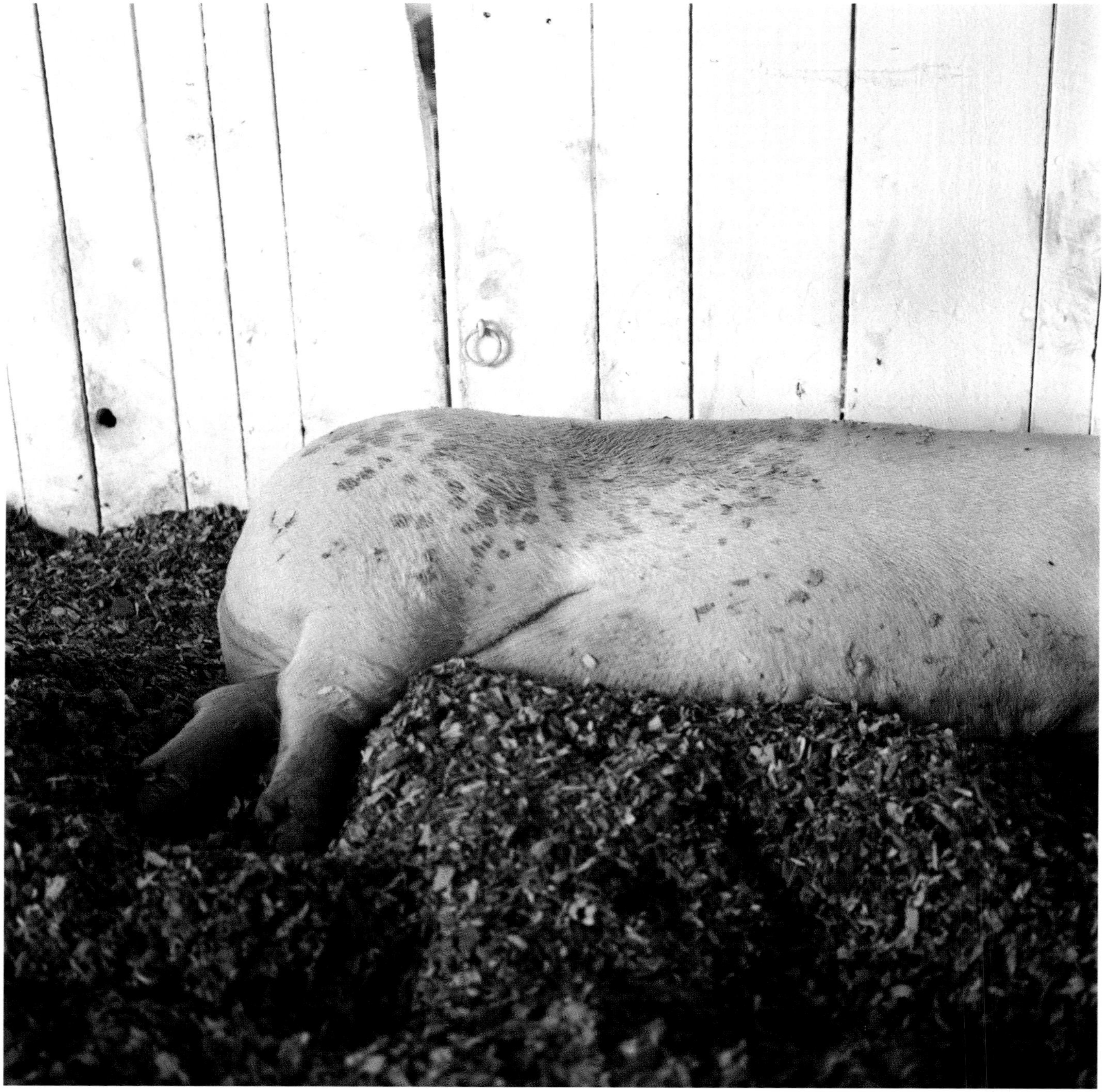

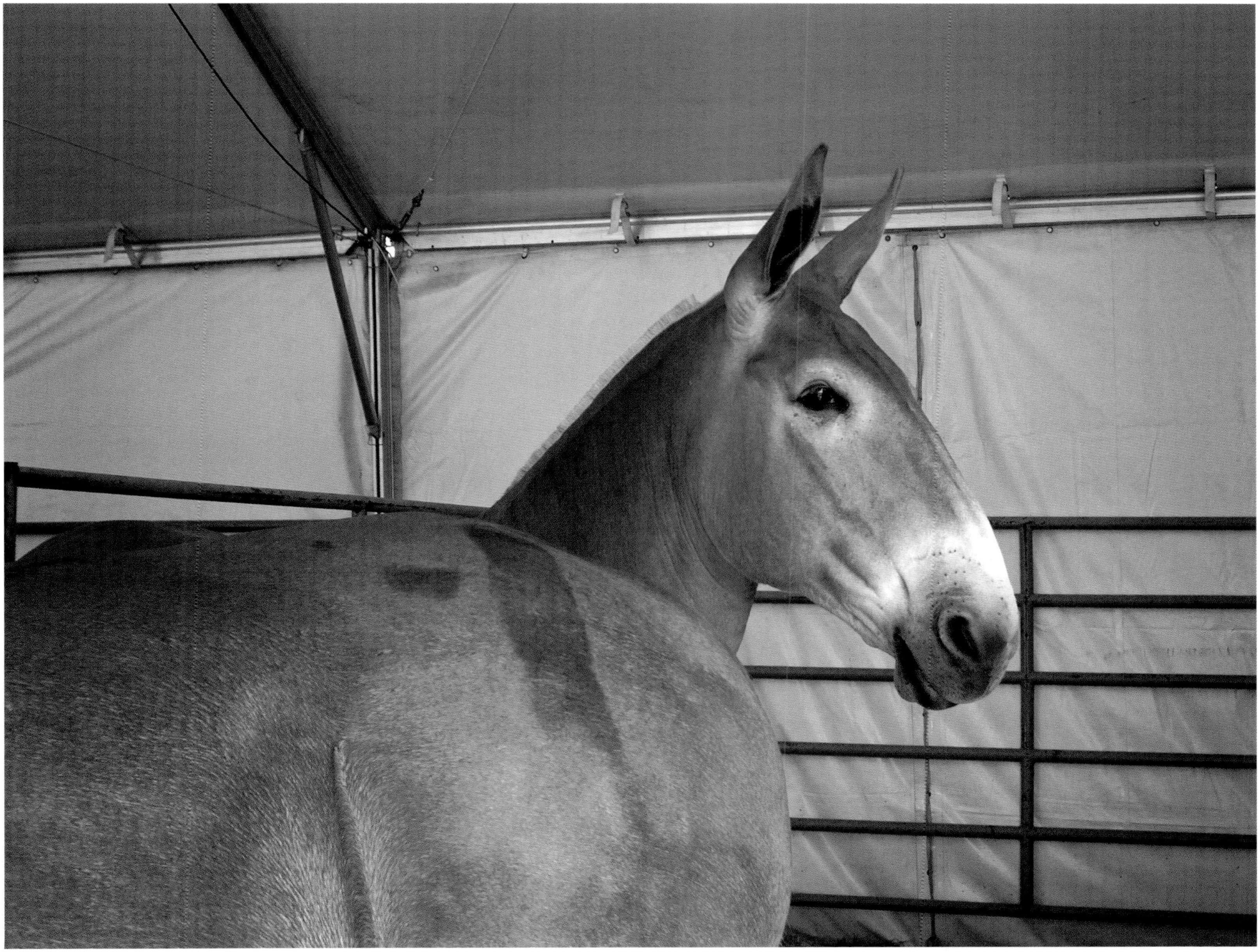

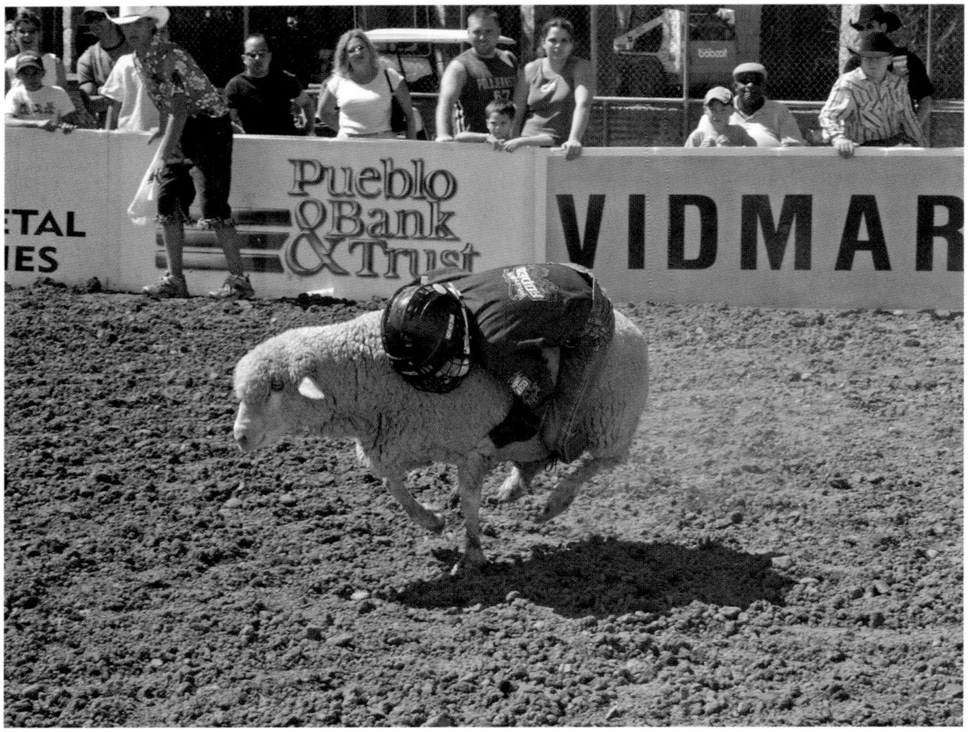
ETAL
IES
Pueblo Bank & Trust
VIDMAR

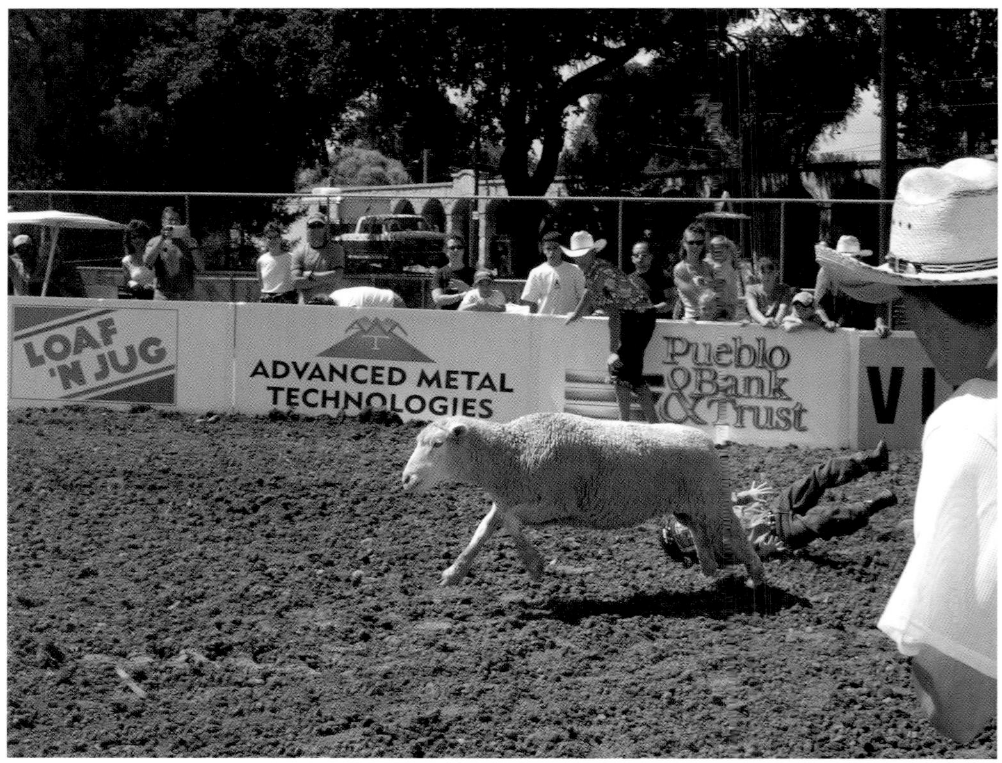
LOAF 'N JUG
ADVANCED METAL TECHNOLOGIES
Pueblo Bank & Trust

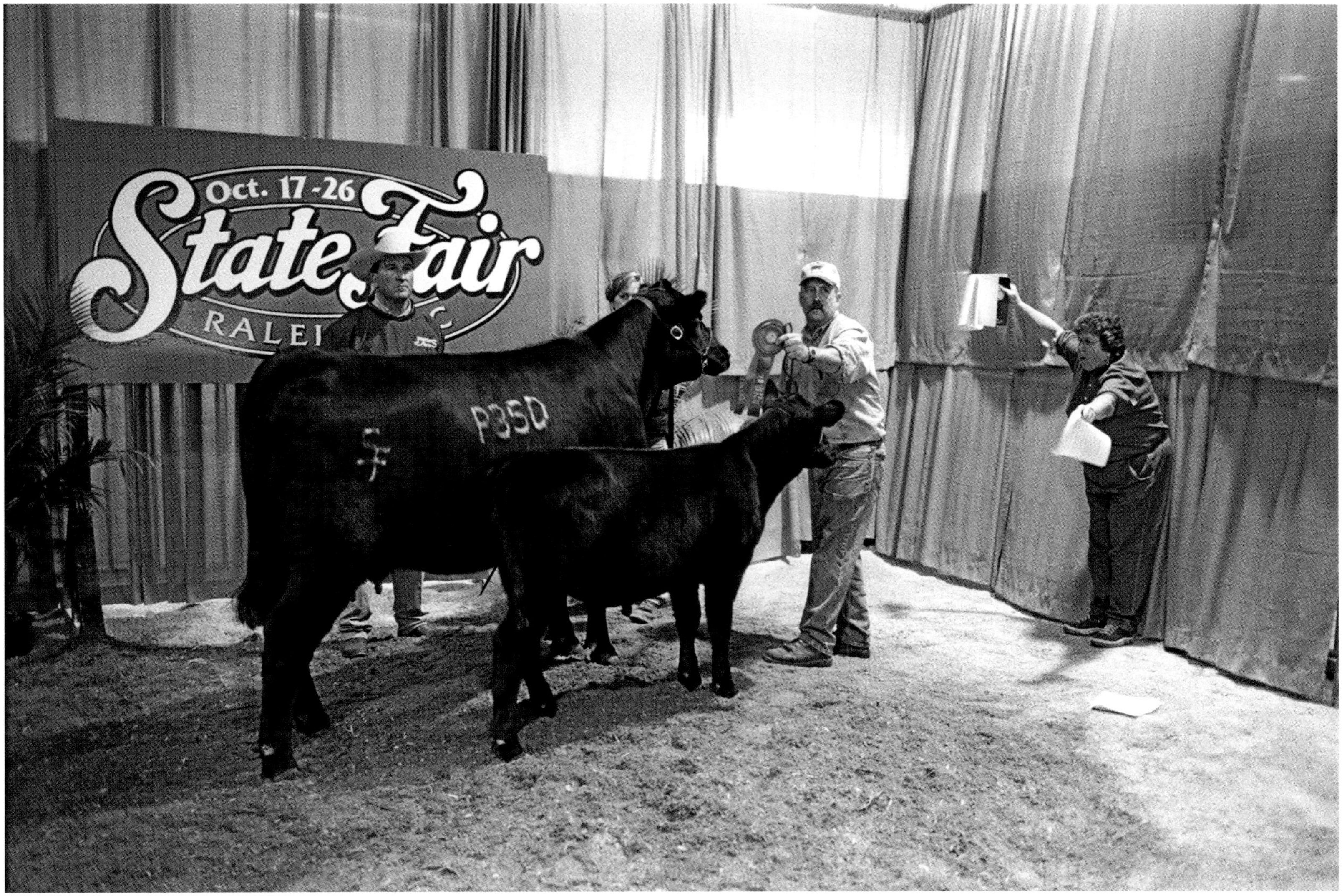
Oct. 17 -26
State Fair
P35D

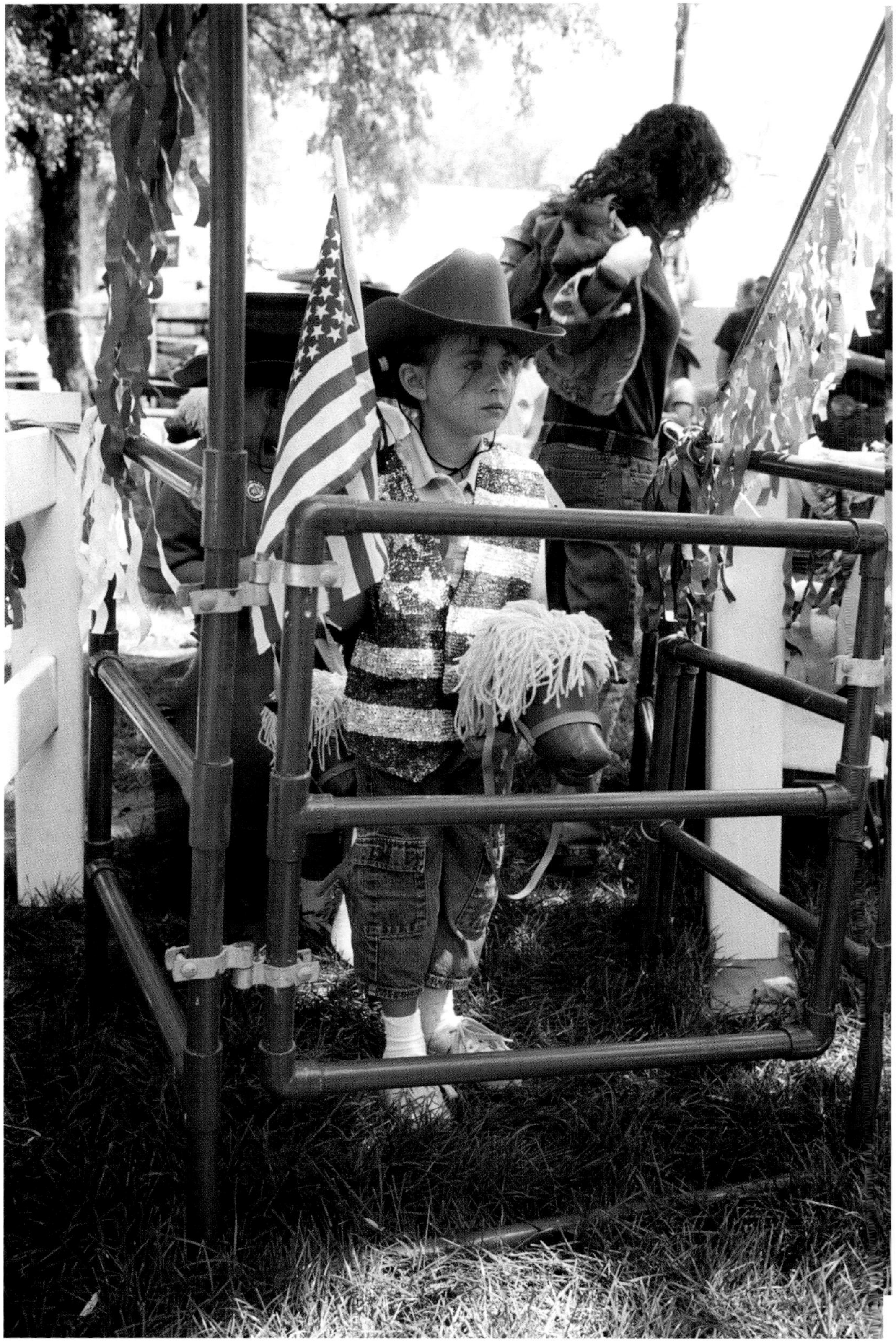

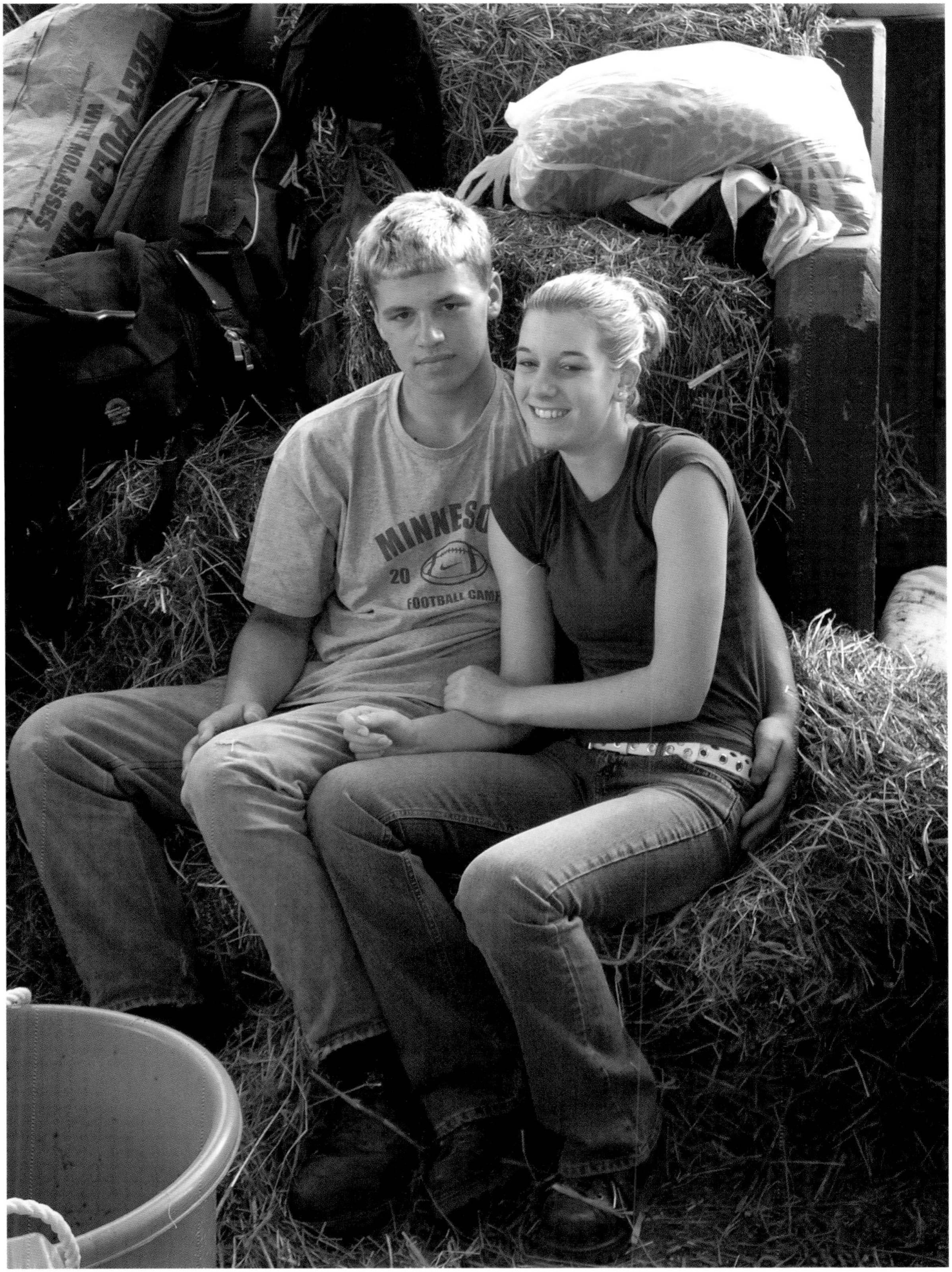
MINNESO
20
FOOTBALL CAMP
WITH MOLASSES

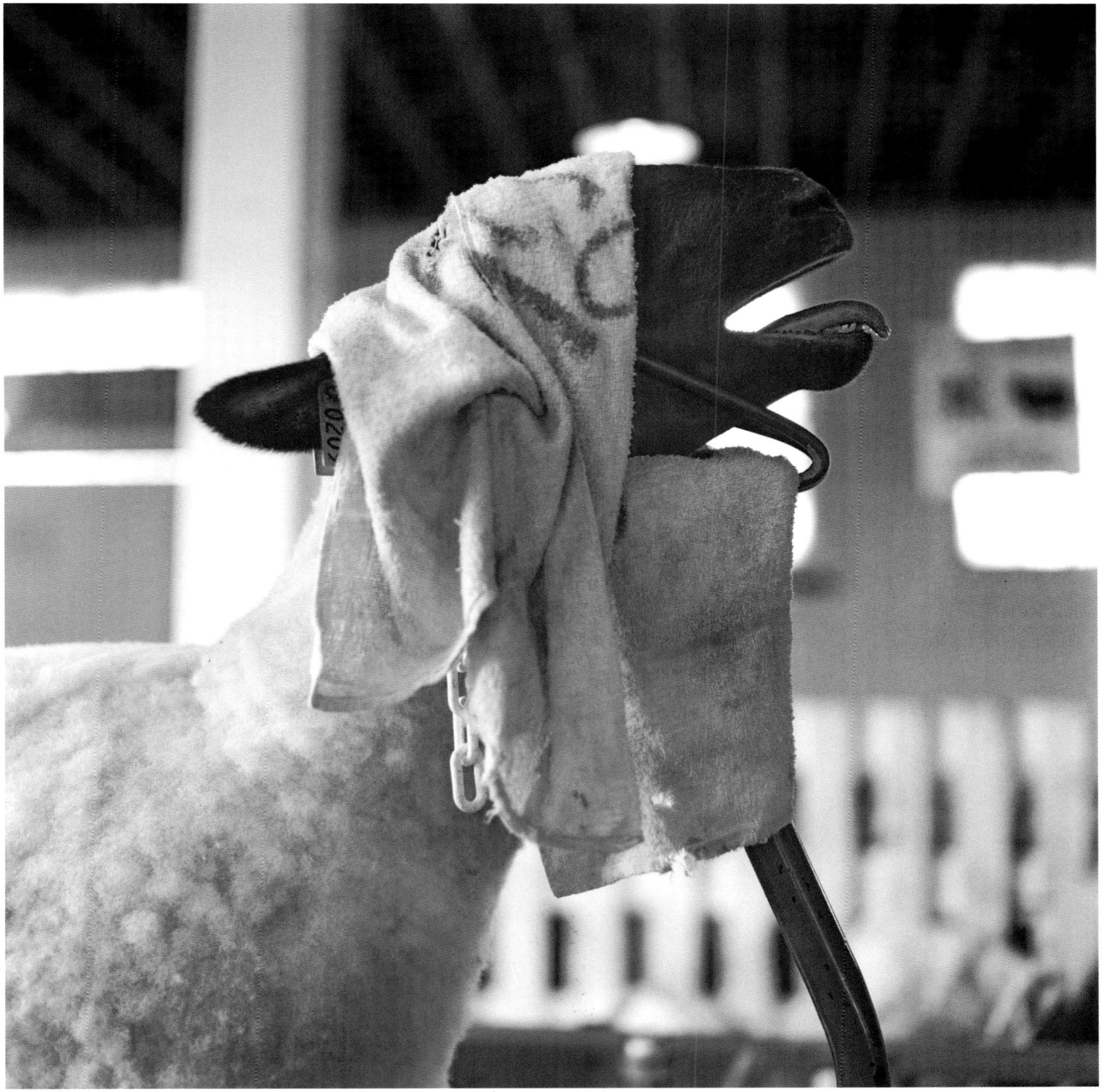

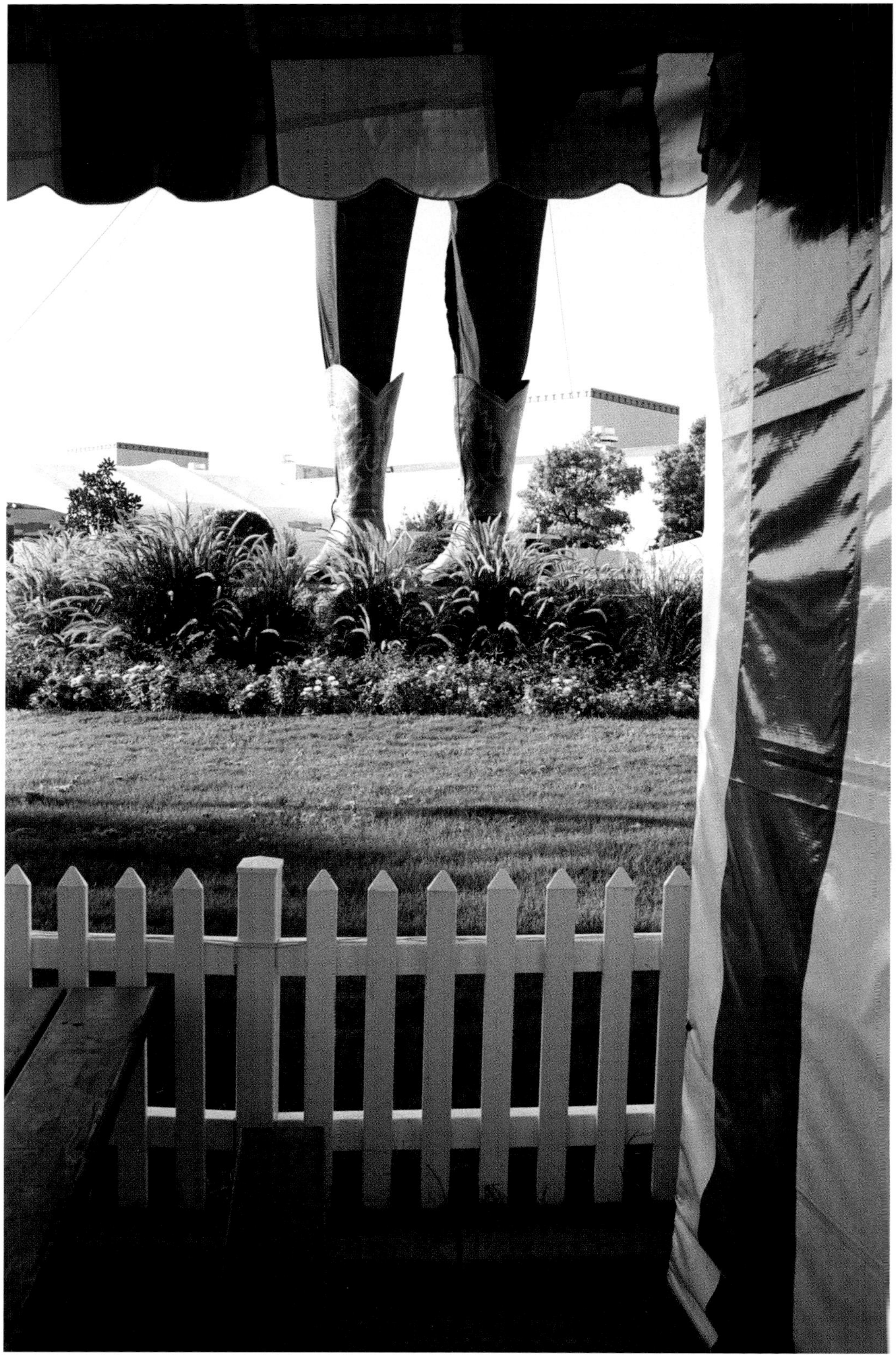

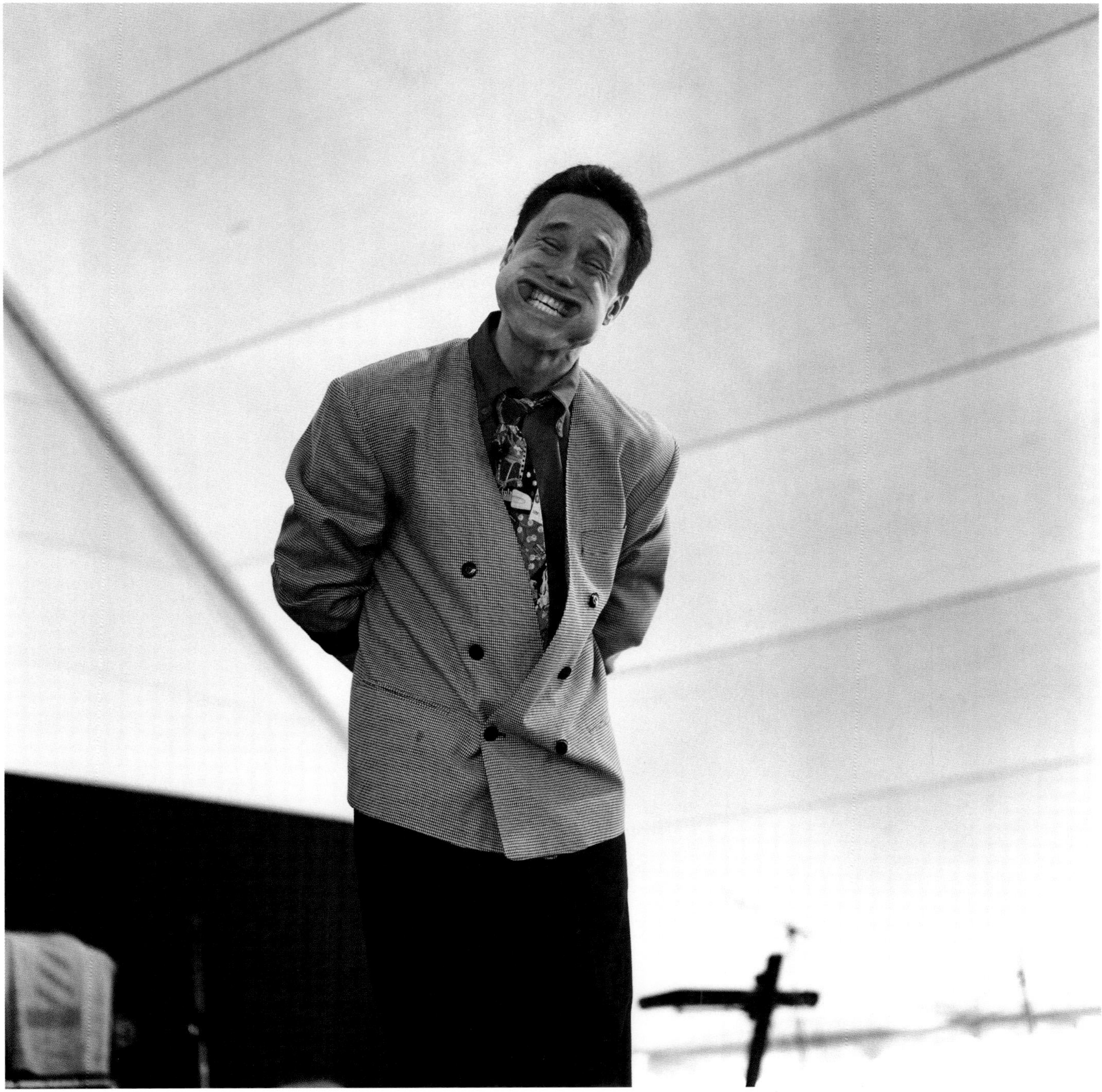

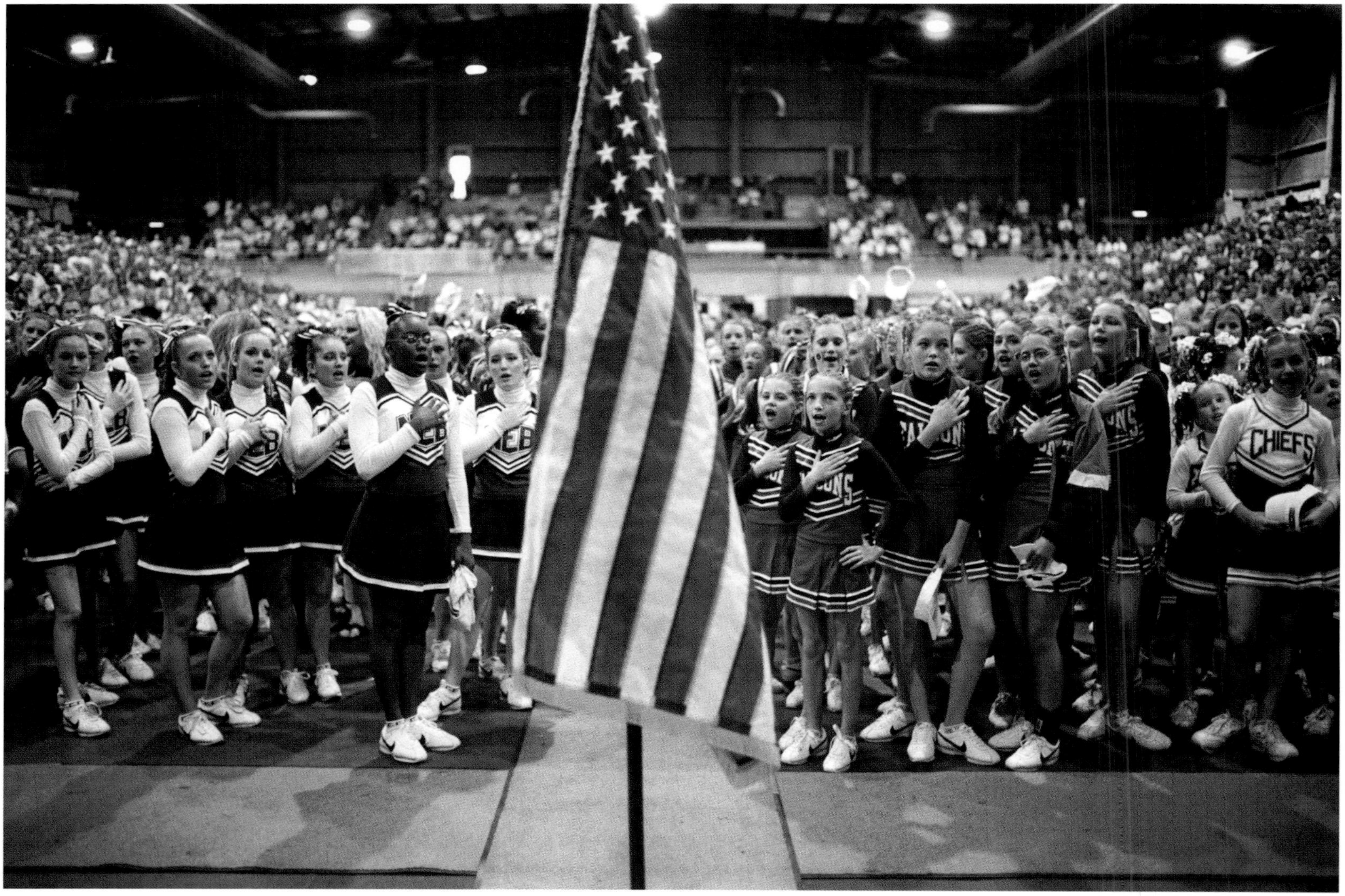
CHIEFS

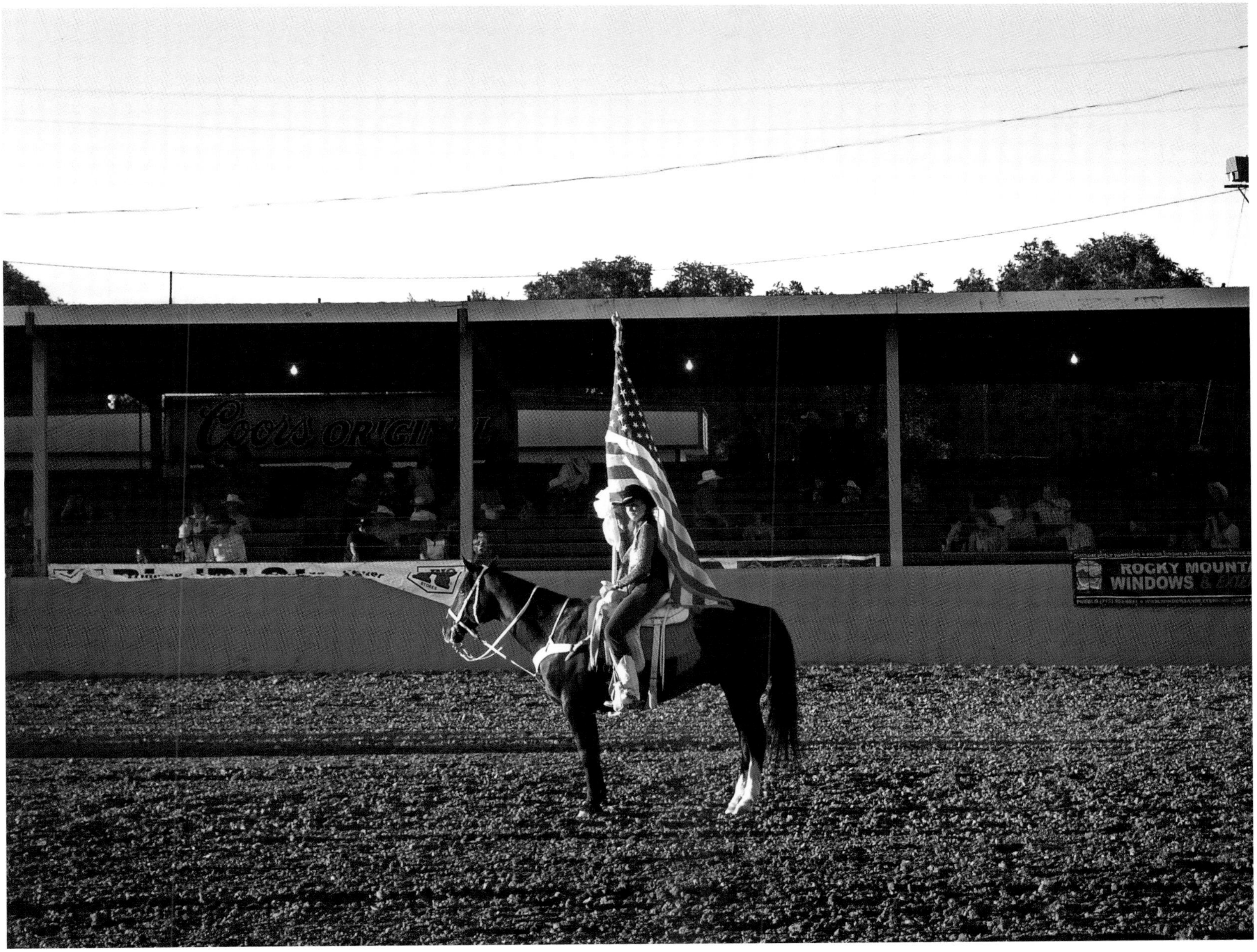
Coors
ROCKY MOUNTA
WINDOWS

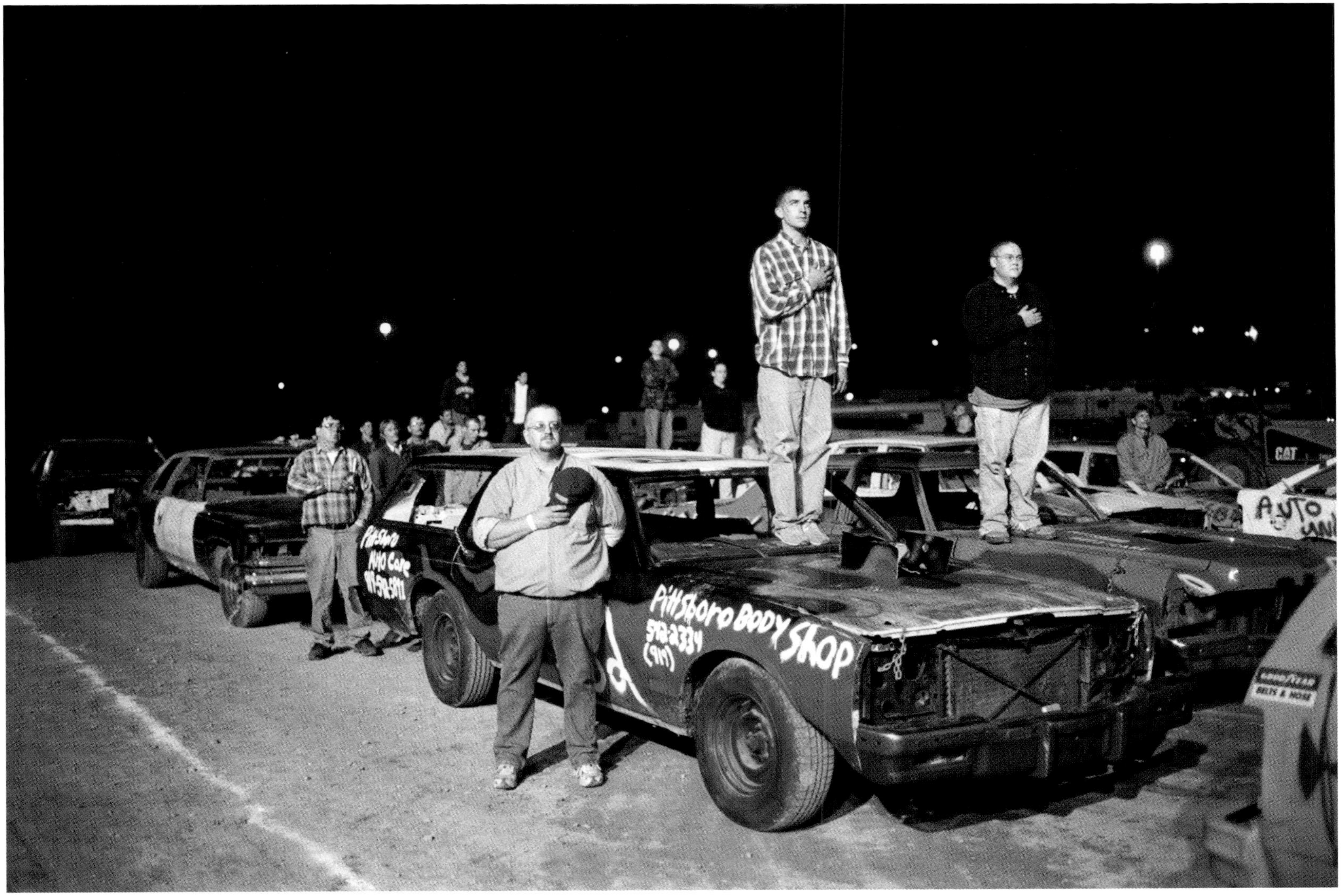
Pittsboro Body Shop
CAT
GOODYEAR
BELTS & HOSE

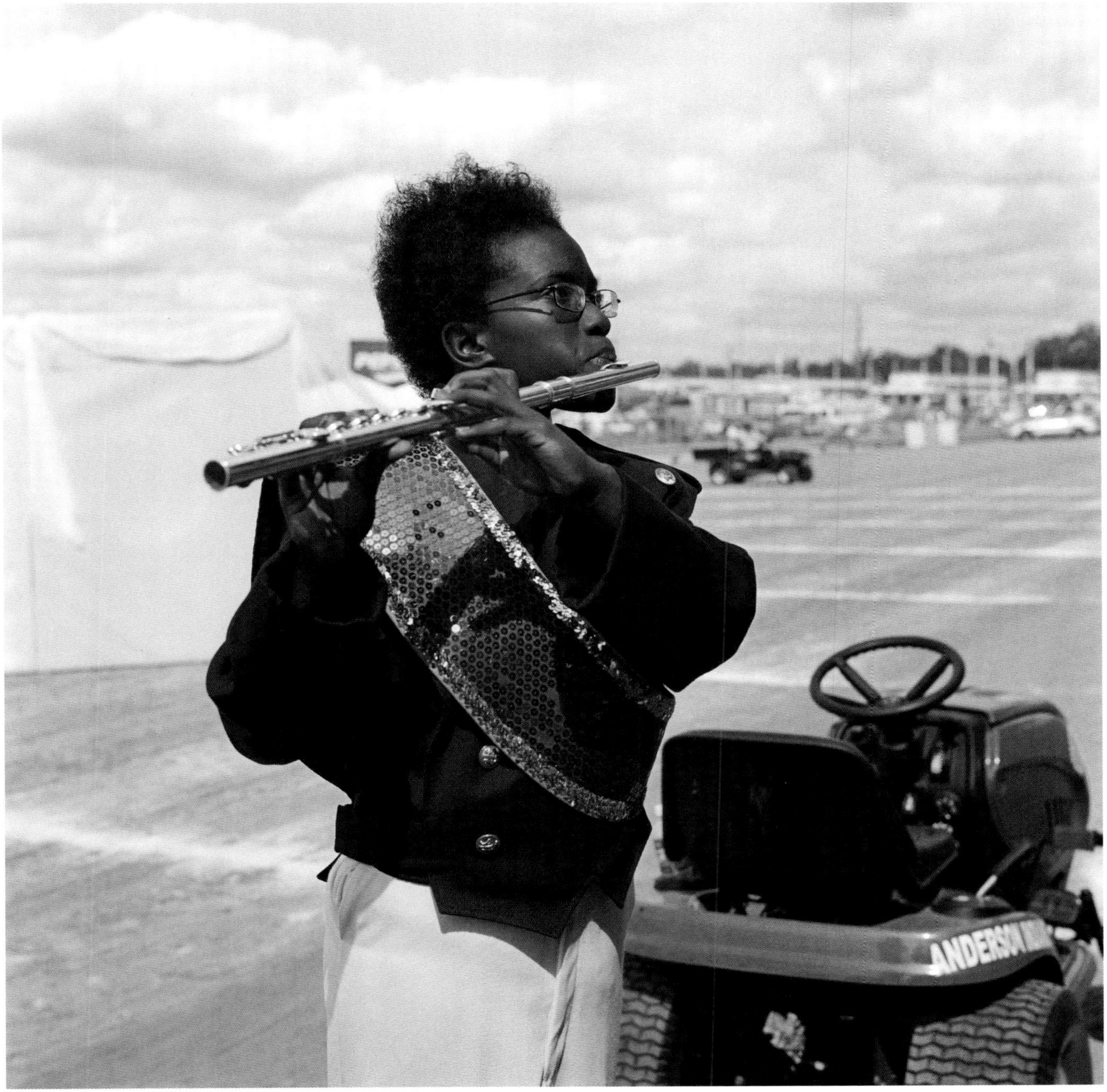
ANDERSON

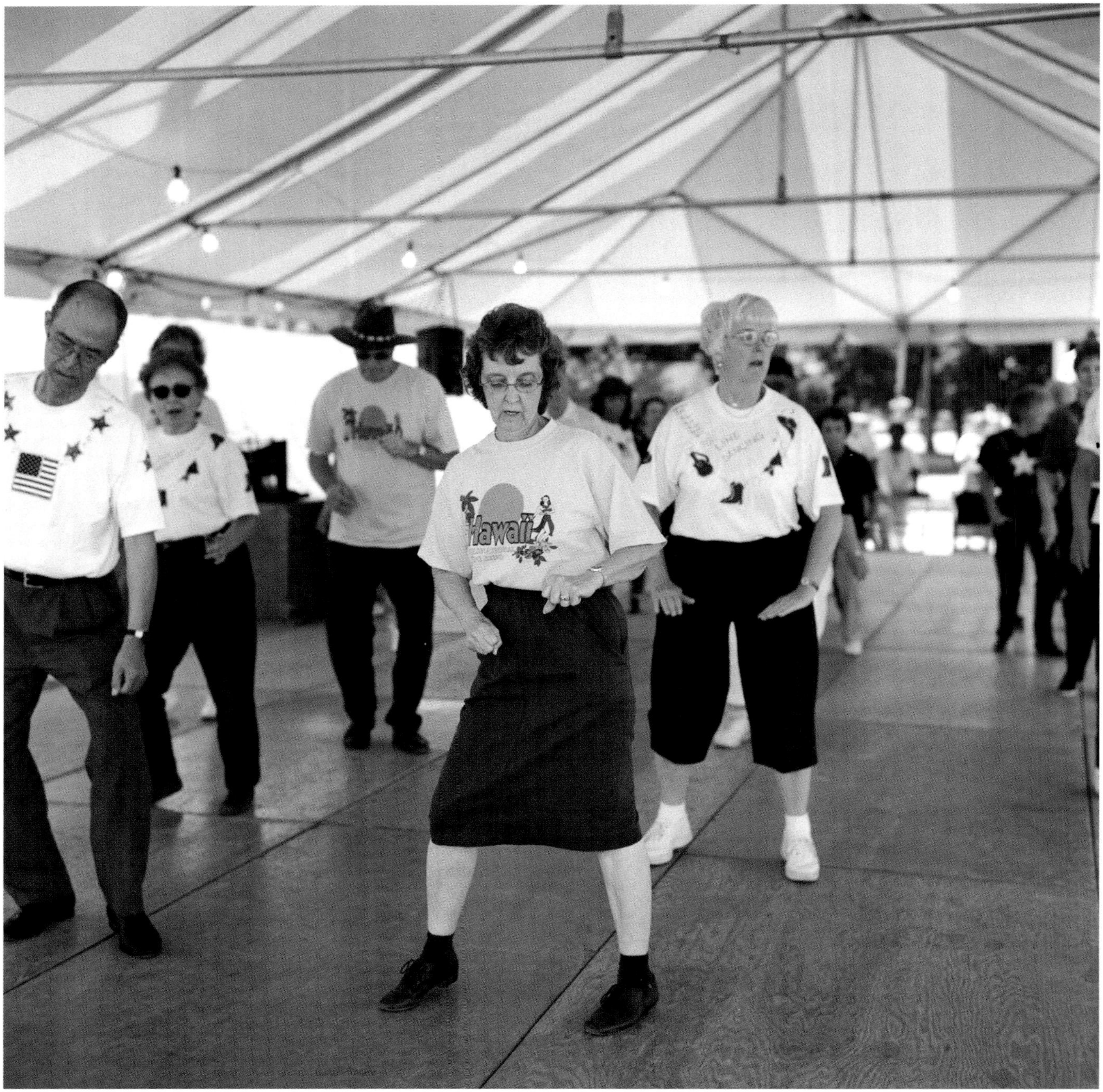
Hawaii

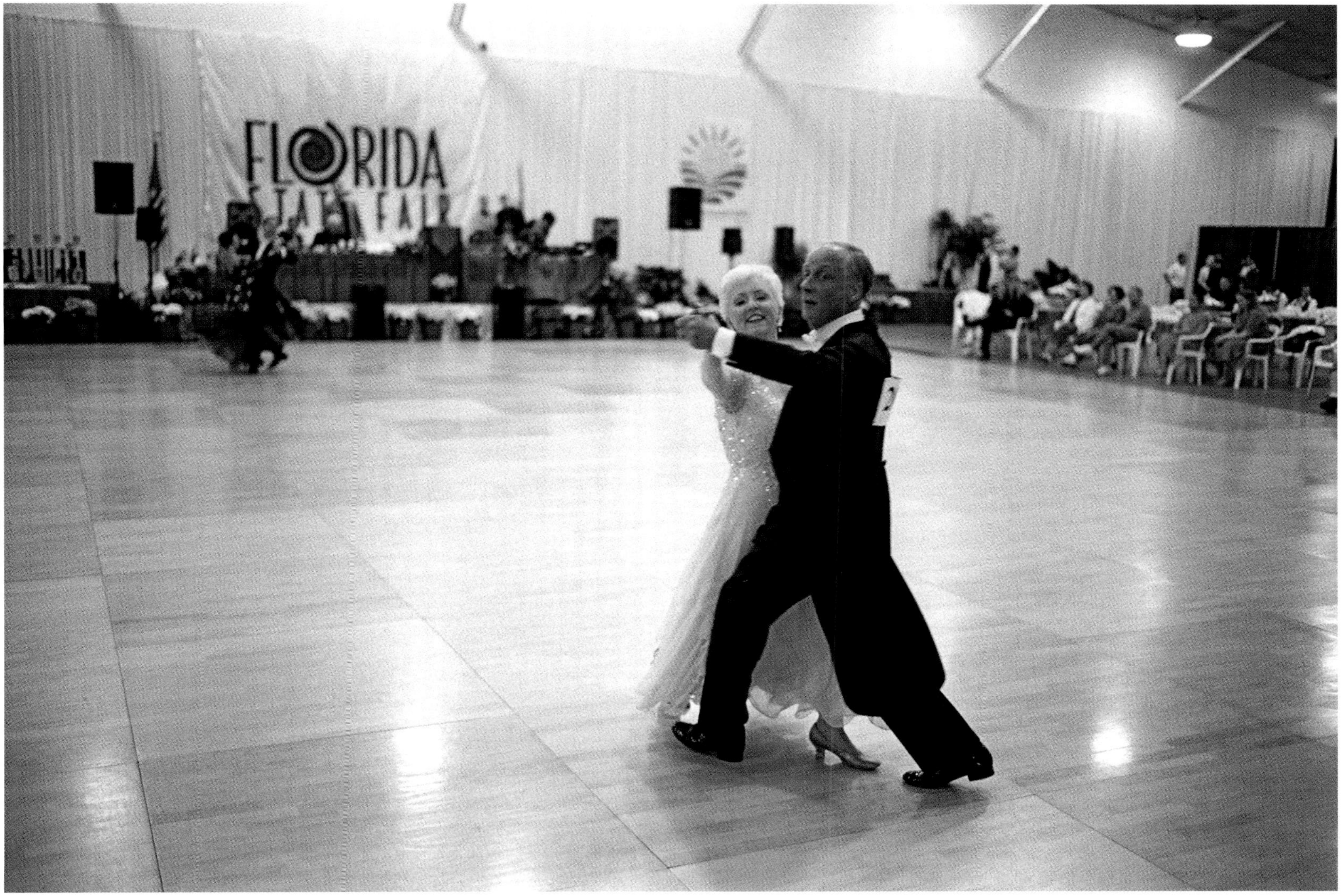
FLORIDA

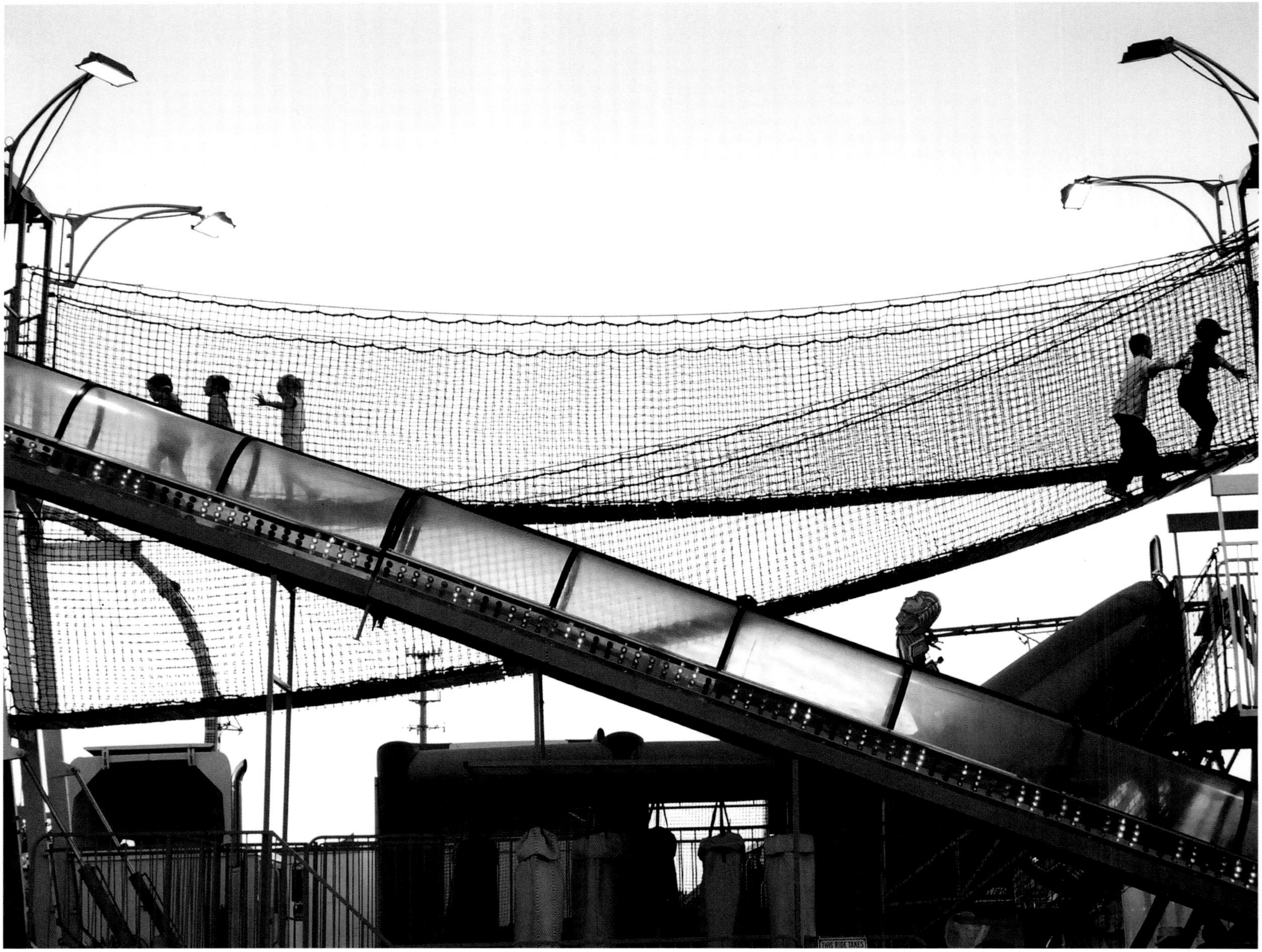
THIS RIDE TAKES

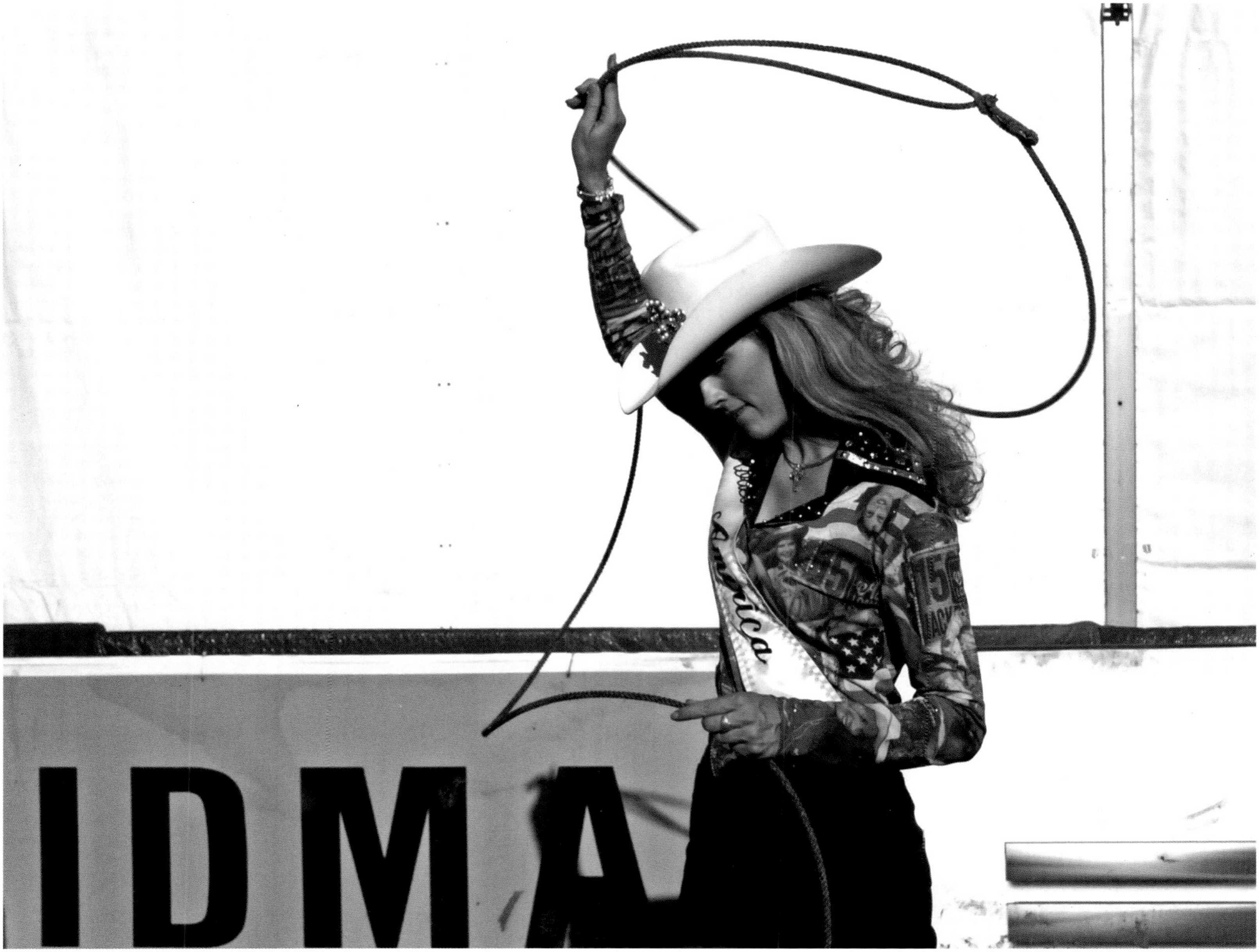
IDMA
America

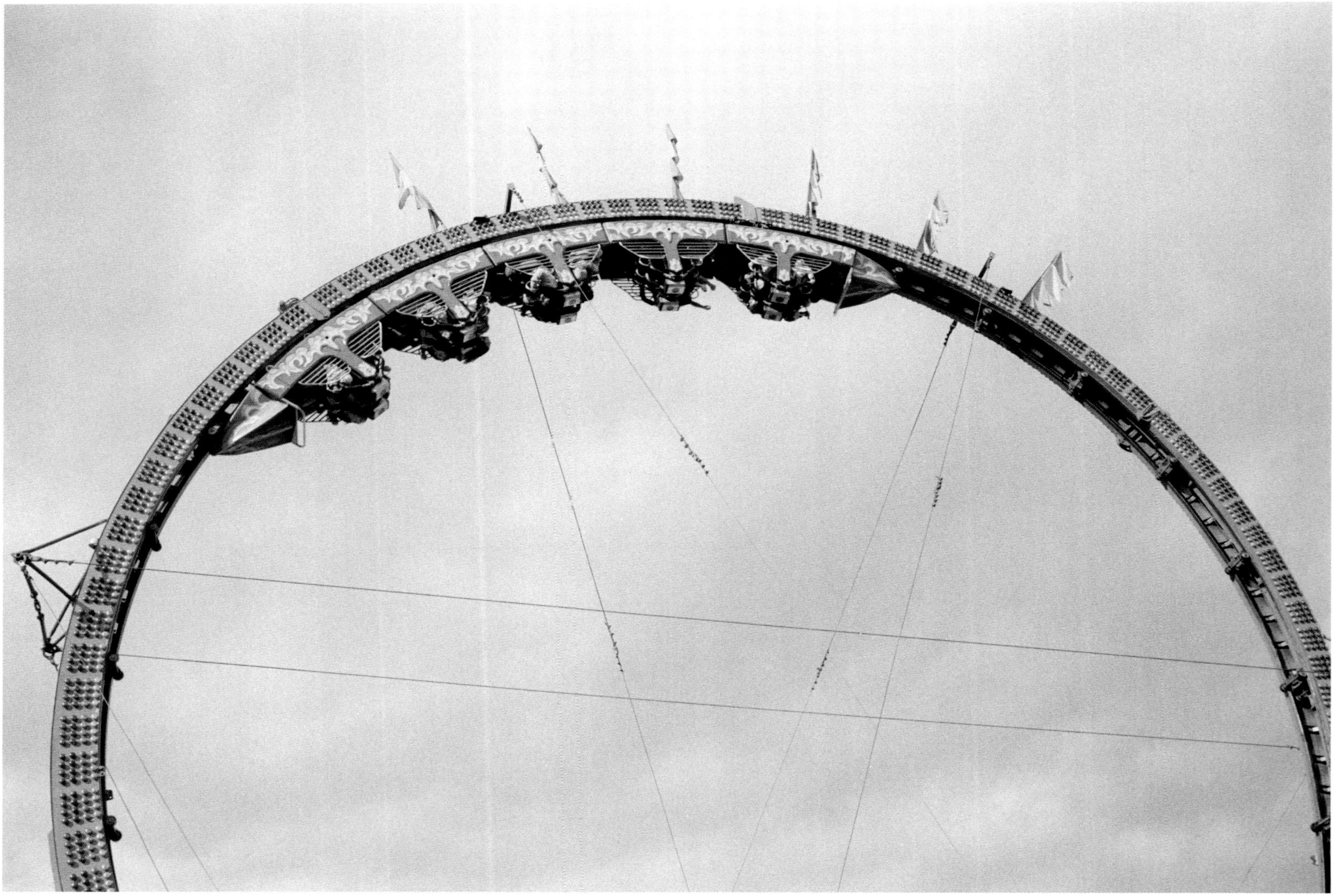

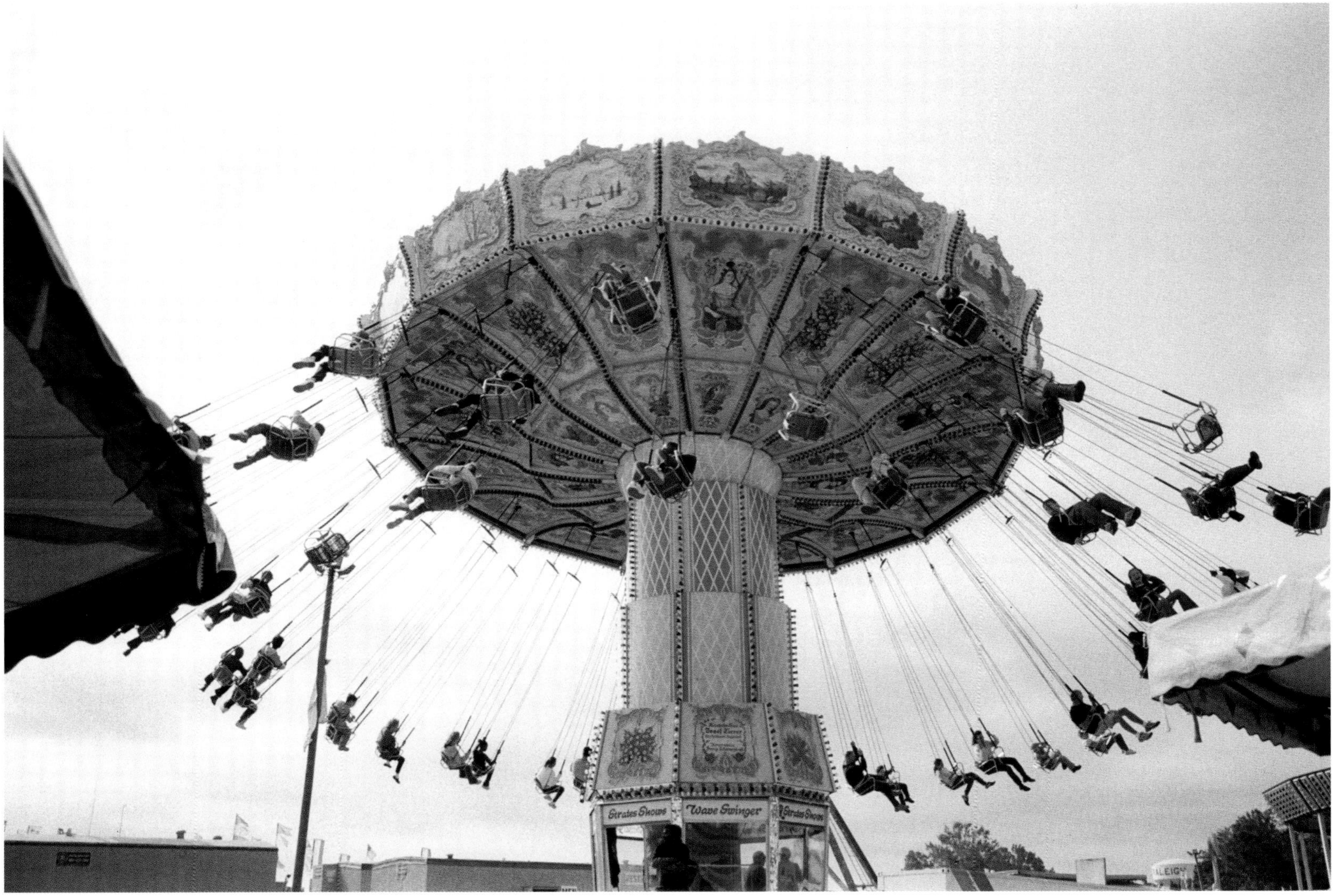
Strates Shows
Wave Swinger
Strates Shows

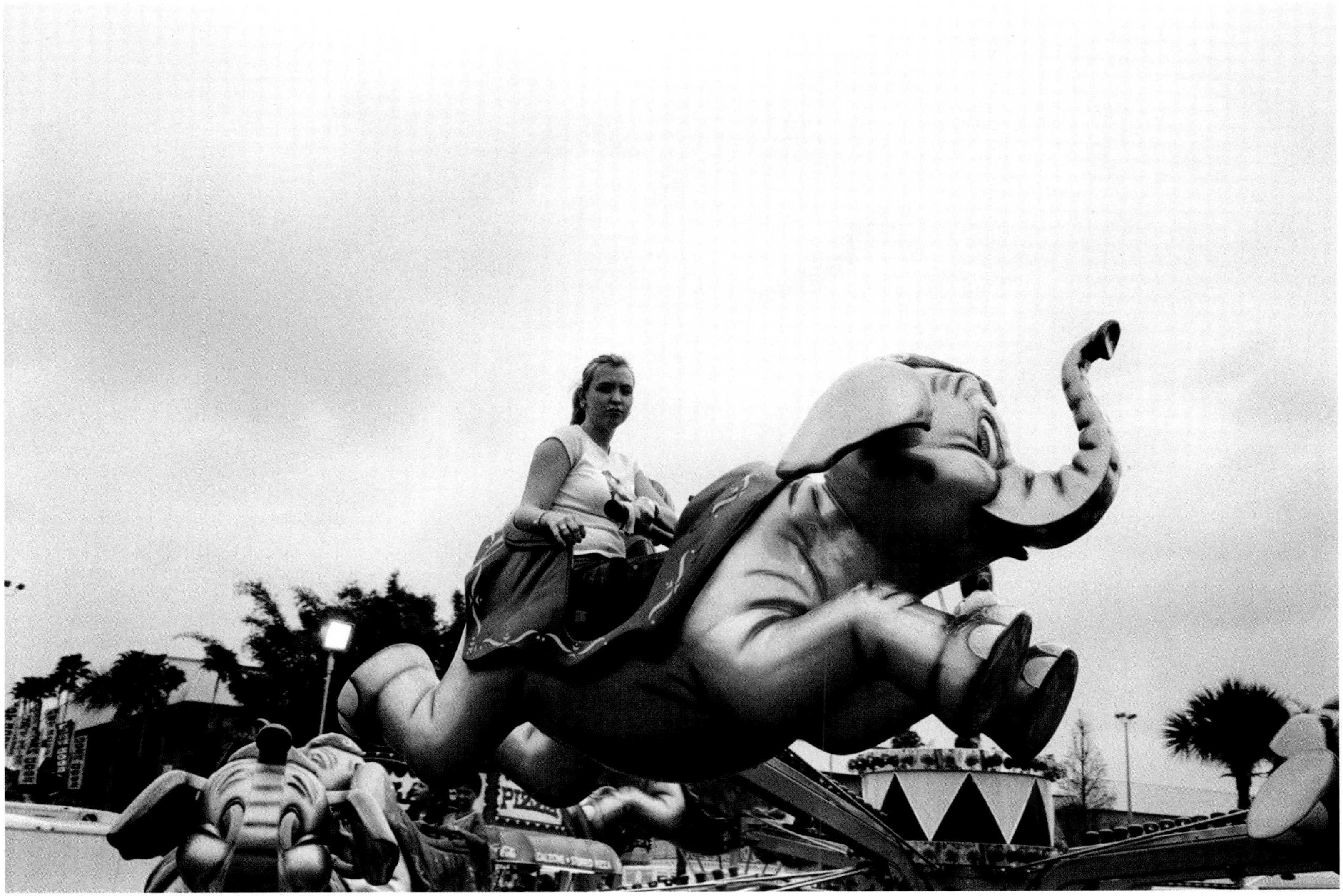

CAPTIONS

17: State Fair of Texas (Dallas), 2004: Standing 52 feet tall and wearing a 75-gallon cowboy hat, "Big Tex" greets visitors to the fairgrounds.

18: Kansas State Fair (Hutchinson), 2004: Fans of rock group R.E.O. Speedwagon peer through grandstand fence during late-afternoon sound check.

19: State Fair of Texas (Dallas), 2004: Food vendor takes a "siesta break" in a patch of shade during a late-afternoon lull in business.

20: Minnesota State Fair (St. Paul), 1977: High school marching band majorettes stand at "parade rest" with trophies after winning statewide competition.

21: Indiana State Fair (Indianapolis), 2004: High school marching band poses for formal group photo after performing in statewide Band Day competition.

22: Colorado State Fair (Pueblo), 2004: Fairgoers pay their respects as the national anthem is played over loudspeakers before the start of the children's mutton bustin' competition.

23: Minnesota State Fair (St. Paul), 2004: Float honoring the heroes of 9/11 moves along parade route during a tribute to firefighters.

24: North Carolina State Fair (Raleigh), 2003: Three police officers on patrol at fairgrounds.

25: California State Fair (Sacramento), 2003: Two police officers on duty at the entrance to fairgrounds.

26: Kansas State Fair (Hutchinson), 2004: "Uncle Sam" performer on stilts entertains fairgoers.

27: Indiana State Fair (Indianapolis), 2004: "Vaudeville" entertainer performs on makeshift stage at Senior Citizens Building.

28: Minnesota State Fair (St. Paul), 2004: Fairgoers take a sit-down break along midway.

29: Indiana State Fair (Indianapolis), 2004: Float riders whoop it up before the start of early-evening parade around fairgrounds.

31: State Fair of Texas (Dallas), 2003: Sunni Cranfill, Miss Texas 2003, signs autographs at the Chevrolet exhibit booth.

32: State Fair of Texas (Dallas), 2003: Clydesdale horse on display at Budweiser beer exhibition area.

33: Florida State Fair (Tampa), 2004: Alligator in wading pool waits to be "wrestled" by the fearless Kachunga.

34: Indiana State Fair (Indianapolis), 2004: Swimming championships float moves along parade route through fairgrounds.

35: Florida State Fair (Tampa), 2004: Life-size "G.I. Joe" rides atop motorcycle during daily parade through fairgrounds.

36: Minnesota State Fair (St. Paul), 2004: University of Minnesota cheerleading squad rehearses before performing along parade route.

37: Indiana State Fair (Indianapolis), 2004: High school marching band performs routine in front of grandstand during Band Day competition.

39: California State Fair (Sacramento), 2003: 4-H Club prizewinners in egg competition proudly display their cartons of perfect eggs.

40: Indiana State Fair (Indianapolis), 2004: Winning bales of hay with prize ribbons attached sit on display in Agriculture Building.

41: Colorado State Fair (Pueblo), 2004: Prizewinning vegetables on display in refrigerated case.

43: Kansas State Fair (Hutchinson), 2004: Arm wrestling competitor prepares for start of match.

44: Minnesota State Fair (St. Paul), 2004: University of Minnesota "spirit squad" in discussion before performing in daily parade.

45: Indiana State Fair (Indianapolis), 2004: State fair queens pose on float before start of early-evening parade.

47: Florida State Fair (Tampa), 2004: Members of cheerleading team relax before competing in statewide championship.

48: Minnesota State Fair (St. Paul), 1977: Two boys sleep next to cows in dairy barn.

49: Ohio State Fair (Columbus), 2004: Father and sons pose with prizewinning Simmental in cattle barn.

50: Colorado State Fair (Pueblo), 2004: Judge ponders decision during hog competition in Livestock Pavilion.

51: Kansas State Fair (Hutchinson), 2004: Competitors await judge's decision for Grand Champion lamb during Grand Drive gala.

52: Ohio State Fair (Columbus), 2004: Winner and runner-up in Ohio Grocers Association's Best Bagger Contest pose with bagging trophies.

53: Ohio State Fair (Columbus), 2004: Two weight lifters in bench press and dead lift competition relax between lifts.

54: Kansas State Fair (Hutchinson), 2004: Pie-tasting judge scrutinizes entries during Pillsbury Refrigerated Pie Crust Contest.

55: Kansas State Fair (Hutchinson), 2004: Contestant describes how beef jerky is made during 4-H demonstration competition.

57: North Carolina State Fair (Raleigh), 2003: Food stands await customers in outlying area of fairgrounds by horse arena.

58: Ohio State Fair (Columbus), 2004: Model of house sits on float before start of parade through fairgrounds.

59: Colorado State Fair (Pueblo), 2004: Elvis statue stands atop shop for collectible pictures and prints.

60: Minnesota State Fair (St. Paul), 2004: Members of IRONJACK & IRONJILL Timber Team compete in logrolling event.

61: Minnesota State Fair (St. Paul), 2004: Fairgoers participate in balancing contest on pails at local TV station's exhibit area.

62: Minnesota State Fair (St. Paul), 2004: Local "bounce" team sends woman member skyward during performance along parade route.

63: Minnesota State Fair (St. Paul), 2004: "Snowflyer" launches high off trampoline during "Big Air Show."

64: Indiana State Fair (Indianapolis), 2004: Riders on the Cliffhanger wait to be released from restraints after coming in for a landing.

65: Minnesota State Fair (St. Paul), 2004: Bungee jumper swings to earth after dropping from crane.

66: Ohio State Fair (Columbus), 2004: Contestant in baton twirling competition performs routine under midday sun for judges and audience.

67: North Carolina State Fair (Raleigh), 2003: Riders come to stop at bottom of "Fun Slide."

68: North Carolina State Fair (Raleigh), 2003: Riders are suspended upside down over fairgrounds during turn on KMG Fire Ball.

69: North Carolina State Fair (Raleigh), 2003: Excited fairgoers experience the thrill of the Top Spin ride.

71: Indiana State Fair (Indianapolis), 2004: 4-H exhibitor awaits rabbit judge during competition in poultry and rabbit barn.

72: Minnesota State Fair (St. Paul), 2004: Two poodles with owner/trainer during break in dog obedience demonstrations.

73: Ohio State Fair (Columbus), 2004: Long-haired Chihuahua, "Bear," waits to receive Outstanding Therapy Dog award.

74: State Fair of Texas (Dallas), 2003: Traditional "carnival barker" plies his trade over loudspeaker system.

75: State Fair of Texas (Dallas), 2003: Win-a-Prize game vendor holds softball while waiting for customers.

76: North Carolina State Fair (Raleigh), 2003: Contestants concentrate on aiming their water guns to break balloons and win prize.

77: Indiana State Fair (Indianapolis), 2004: Young competitor gets ready to go for distance in watermelon seed spitting contest.

79: State Fair of Texas (Dallas), 2003: Competitor in State Fair Corny Dog Contest wolfs down another corny dog in allotted time.

80: State Fair of Texas (Dallas), 2003: Young boy in fatigues with goat outfitted as "tank" waits to enter ring in animal costume competition.

81: Florida State Fair (Tampa), 2004: Young boy tries out anti-aircraft gun at U.S. Army exhibition.

83: California State Fair (Sacramento), 2003: Life-size cardboard poster of CHP officer, with cutout in face area, awaits next person to pose at California Highway Patrol booth.

84: Colorado State Fair (Pueblo), 2004: Prospective customers try out "Spinal Exerciser" for back pain relief and weight loss.

85: Minnesota State Fair (St. Paul), 2004: Karaoke singers perform at Farmers Union Agency booth.

86: Minnesota State Fair (St. Paul), 2004: Minnesota Veterinary Medical Association doctor explains procedure to audience during live spaying operation on cat at the State Fair Surgery Suite.

87: Kansas State Fair (Hutchinson), 2004: Woman examines butter sculpture made from 1,000 pounds of butter on display inside refrigerated container.

88: Indiana State Fair (Indianapolis), 2004: 4-H Club chorus performs at opening ceremonies.

89: Minnesota State Fair (St. Paul), 2004: Aquatic performers wave to audience before beginning "High Dive Show" in amphitheater.

90: Minnesota State Fair (St. Paul), 2004: Participants enjoy interactive virtual reality theater presentation of "VR Gone Wild."

91: Colorado State Fair (Pueblo), 2004: Imperial characters of *Star Wars* mingle with children at Fountain Plaza.

92: State Fair of Texas (Dallas), 2003: Couple poses with fake trophy against raceway backdrop during promotional photo op at Dodge cars exhibition.

93: Colorado State Fair (Pueblo), 2004: Official "Spiderman" impersonator heads out of dressing room toward stage for Meet & Greet event with children.

94: State Fair of Virginia (Richmond), 1992: Hooded sheep rest in pen before start of judging.

95: State Fair of Texas (Dallas), 2004: Young boy and goat, dressed up as pirates, get ready to enter ring for animal costume competition.

96: State Fair of Virginia (Richmond), 1992: Torso of large hog at rest in holding pen.

97: Colorado State Fair (Pueblo), 2004: Belgian draft mules wait in stall before rodeo parade.

98: Colorado State Fair (Pueblo), 2004: Child riders try to hang on for the most seconds in mutton bustin' competition.

99: North Carolina State Fair (Raleigh), 2003: Photo "wrangler" waves papers to get attention of prizewinning cattle during official photo op.

100: Colorado State Fair (Pueblo), 2004: Young participant in Cowboys & Kids Pee-Wee Stampede waits to enter ring on her broomstick pony.

101: Minnesota State Fair (St. Paul), 2004: Young couple relaxes in cattle barn during break in judging.

103: State Fair of Virginia (Richmond), 1992: Sheep in restrainer bleats while being trimmed for show ring.

104: State Fair of Texas (Dallas), 2003: Size 70 boots of "Big Tex" stand 7′7″ tall on fairgrounds.

105: State Fair of Virginia (Richmond), 1992: Juggler with balls in mouth mugs for audience during performance.

106: Florida State Fair (Tampa), 2004: Members of cheerleading teams competing in statewide championship sing along while national anthem is played.

107: Colorado State Fair (Pueblo), 2004: Lone color guard on horseback waits for start of opening ceremonies at evening rodeo.

109: North Carolina State Fair (Raleigh), 2003: Participants in demolition derby listen as national anthem is played over grandstand PA system before start of event.

111: Indiana State Fair (Indianapolis), 2004: High school marching band flute player performs during Band Day statewide competition.

112: Indiana State Fair (Indianapolis), 2004: Line dancing group, age 55 and older, gives performance on Dance for Health stage.

113: Florida State Fair (Tampa), 2004: Couple performs during final of ballroom dancing competition.

114: Colorado State Fair (Pueblo), 2004: Children climb walkways of fun house along midway at dusk.

115: Colorado State Fair (Pueblo), 2004: Miss Rodeo America 2004, Darci Robertson, warms up for lassoing demonstration with children.

116: North Carolina State Fair (Raleigh), 2003: Enthusiasts wave arms while upside down on the Fire Ball ride.

117: North Carolina State Fair (Raleigh), 2003: Riders get bird's-eye view of fairgrounds on the Wave Swinger.

119: Florida State Fair (Tampa), 2004: Woman circles solo on "Dumbo the Elephant" ride.